CHAINSAW SAVVY

Published by Morgan & Morgan, Inc.
145 Palisade Street
Dobbs Ferry, New York 10522

International Standard Book Number 87100-187-X Softcover
International Standard Book Number 87100-188-8 Hardcover

Library of Congress Catalog Card Number 82-061035

Cover design: John Alcorn
Cover photo: Liliane DeCock

Printed in U.S.A.

Cover printed by Morgan Press Incorporated, Dobbs Ferry, N.Y.

CHAINSAW SAVVY

A COMPLETE GUIDE

By Neil Soderstrom

Drawings by Richard J. Meyer

Photos by Neil Soderstrom unless
otherwise credited

MORGAN & MORGAN
145 Palisade St.
Dobbs Ferry, N.Y. 10522

Publisher's Note

To the memory of Robert C. Le Clair,
a teacher who knew the value of encouragement

Contents

PART I. SAFE, EFFICIENT SAW USE

Can a Book Help?

On a TV talk show, I once suggested that a chain-saw is the most dangerous tool in common use. Shortly after the show, a chainsaw representative reproached me, saying, "What you said about the dangers of chainsaws really shocked me. It cost the industry *at least* 2,000 sales!"

His words shocked *me*. Wasn't he advising his staff to sell safety along with saws? Had years of struggle with sales quotas warped his outlook that much? Did he lie awake nights calculating his company's share of the market with no thought to its share of the injuries?

The fellow must have tuned out, though his TV played on, the moment I linked chainsaws with danger. For, on the air, I'd gone on to say that safe-ty features on better saws can help prevent some injuries but that nearly all injuries result from user errors—violations of safety instructions in owner's manuals. My point was that saw injuries could be nearly eliminated if people made an effort to operate saws correctly.

It's a lot easier to sell chainsaws than chainsaw safety. Marketing strategies that produce high-volume sales usually stress bargain prices—not safety features, not safety instructions, and not the daily maintenance needed to keep a saw running as safely as it's designed to. In fact, some chainsaw companies now market their saws through stores where customers receive no saw-use demonstra-tions whatever. Customers simply pick up their boxed bargains and go home to assemble them. If it's an electric saw, the new owner plugs it in and assaults his woodpile. If it's a gas saw, he must first attempt to puzzle out the carburetor settings before he can make firewood. Some new owners read beyond assembly and start-up instructions and go on to study safety instructions as well. Many do not. And many suffer mutilating injuries.

The cheapest saws, those that sell in highest quantities, might best be considered *disposable* because costs for repairs can approach the pur-chase price. The cheapest saws may not be designed to run 100 hours and have traditionally lacked rudimentary anti-kickback safety features. They are noisy. They vibrate excessively. Their left and right handles are too close together to allow good control. An ace mechanic once characterized the cheap saws to me as "designed to self-destruct." Yet these cheap saws sell well and so will probably continue to abound in the market place.

The irony here is that it's not necessarily cheap saws themselves that are to blame for accidents. They're merely part of the cheap approach to marketing, retail sales, and owner education. And since the new owner doesn't spend terribly much for his saw, he may not feel compelled to study his owner's manual thoroughly—especially those parts devoted to safe use. Truly, in the right hands the cheapest saws with the fewest safety features can still perform efficiently and safely. In the wrong hands, even the finest saws with the best safety features bring sorrow.

Manufacturers contend that even the most sophisticated saw safety features can still offer only limited protection—that a chainsaw must be inherently dangerous if it is to be an efficient wood-cutting tool. This argument seems legitimate to me. As long as chain is used to cut wood, engineers will have trouble designing saws that are significantly safer than the best ones are today—unless efficiency is sacrificed. Yet, someday a technological revolution will occur. It will perhaps relegate chainsaws to history. That is, woodcutters of the future may not be using chainsaws at all. They may be slicing almost silently through massive logs by means of laser beams or wands right out of science fiction.

But all that's for future shock. Right now we've got chainsaws. They're capable of sweet-running productivity as well as instant horror. What can be done to make their use safer? Plenty. We could make chainsaws less accessible. Maybe chainsaws shouldn't be sold to just anyone who's got the dough for them. And maybe chainsaws shouldn't be operated by just anyone who can lift them to a log. If user errors account for nearly all chainsaw injuries, why not require that would-be users first receive safety training and then pass proficiency tests? Such training and tests have proven successful for hunters, drivers, and scuba divers. Why not for chainsaw users? In addition to learning safe methods, trainees would automatically be learning efficient methods too. If trainees also received lessons on chain sharpening and saw maintenance, they would eventually save themselves the price of instruction by depending less on service shops. Alas, that day is a long way off.

Until that day, maybe this book will help reduce the annual carnage and promote efficient saw use. This book is not intended to substitute for your saw manual because owner's manuals can address the maintenance and operating considerations peculiar to specific saw models. Yet, if you first study your owner's manual, this book should be a useful supplement. It might scare the hell out of you at times, but a good scare sure beats a bad cut when you have the choice.

For the record, I should add that many manufacturers submitted their products to me for testing and photos. These submissions came with no strings attached. Also, various industry experts consented to review my manuscript and offer suggestions. Although I am grateful for the product submissions and for the suggestions, my gratitude has not influenced my evaluations of any saw features or accessories. In fact, some manufacturers that declined to participate entirely will find their products roundly applauded simply because the products are good ones. Yet, since nearly all good products (cooperating manufacturer or not) have negative as well as overriding positive features, you'll find these negatives mentioned too. I hope my independence shines through.

Neil Soderstrom

Acknowledgments

My thanks go to the following people: To Doug Morgan, who as part-time chainsaw dealer suggested the need for this book and then as full-time president of Morgan & Morgan, Inc., offered to publish it. To Daniel Tilton, instructor in Swedish logging methods for Tilton Equipment Company, who demonstrated his considerable skills for my camera and contributed good stuff to the manuscript. To Mike Isser, who arranged for sharp-eyed reviews of the manuscript by two experts with Homelite: Eddie Turner, Forest Product Service Manager, and Vince Morabit, Director of Product Safety. To Gordon Hodson, Product Service Manager for Oregon Saw Chain, who strengthened the chain and bar manuscript and granted permission to reprint Oregon drawings. To Brian Lepine, northeastern Territory Manager of Carlton Company's Saw Chain Division, who offered advice on some of the photos and manuscript. To world-class competition lumberjacks who granted interviews that figured in large and small ways here: Tom and Alex Bildeaux, Dave Geer, Sven and Ron Johnson, Jim Taylor, Ron Hartill, and Roy Booth, Sr. and Jr. To Tom Kuck, a chainsaw dealer who loaned his shop and mechanics for photos; assisted with the maintenance manuscript; and loosed Joe Kincart to expand the troubleshooting chart. To Horace Walters, who deserves a small-engine PhD. To Sid Latham, master photographer, who showed me a good set of lights. And finally to my wife Hannelore and son Nikolai, for all the rest.

N.S.

1

Injury Perspectives

There are roughly 18 million chainsaws in use in the United States. Each year these saws figure directly in about 100,000 injuries requiring medical attention. That averages 274 injuries a day, 365 days a year.

Back in 1976 the Consumer Product Safety Commission studied chainsaw injuries treated in hospital emergency rooms. As to severity of injury, the Commission found that 97 percent of the victims were treated and released. Males made up 95 percent of the victims. Females injured had been working both as sawyers and helpers. Just over 90 percent of all chainsaw injuries involved lacerations. The remainder included amputations, avulsions, fractures, contusions, and abrasions. Percent of injuries by body part were as follows:

	Percent
Finger or hand*	40
Lower arm	10
Lower leg	27
Upper leg	6
Face, head, or neck	6
Other	11

Some of these injuries occurred when the chain was at rest, while the engine was idling or shut off—and during sharpening.

The Commission also studied causes for 30 chainsaw-related deaths that occurred over a four-year period. As you might expect, two thirds of the deaths resulted from neck injuries. In seven cases, the victim fell onto the chainsaw. In five cases, neck-lacerating kickback occurred. Four deaths resulted from trees falling onto the victim or onto his saw which in turn cut him. Other deaths resulted from electrocution, carbon monoxide poisoning, and unspecified hazards.

For many years, kickback was blamed for over 30 percent of all injuries. Kickback is usually thought of as the violent rotational whipping of the saw bar at the operator. This can happen when the chain momentarily hangs up when it hits resistance as it arcs around the upper quadrant of the bar nose. In a second kind of kickback, the saw engine is pushed backward when the moving chain along the topside of the bar hits resistance or becomes pinched in the kerf. This push is sometimes called linear kickback.

By the late 1970s, many saws were equipped with anti-kick and reduced-kick devices. These began to lower the percentages of injury primarily due to kickback. Some of these devices include bar-nose guards, left handguards, chain brakes, banana-nose bars, and "new generation" chain that produces less kicking energy when it does kick. We'll cover these devices more fully in the chapter on kickback.

Anyway, by 1978 the Commission attributed only 23 percent of chainsaw injury to verifiable kickback. Yet, pointing to studies made in Sweden and Canada, the Commission contends that kickback injury could be reduced much further if all saws were fitted with devices that would eliminate or reduce the incidence and severity of kicks.

Aside from design features, some U.S. manufacturers and logging interests argue that a chainsaw must remain inherently dangerous if it is to be an efficient wood-cutting tool. People in this camp contend that education programs could help reduce injuries greatly. No doubt, education could achieve injury reductions. In the accompanying graph, adapted from a Commission report, it seems that injuries are mainly attributable to user errors rather than design inadequacies. For example, in one Commission study of kickbacks, it was found that 25 percent of victims had been operating the saw one-handed, which is a taboo in any owner's manual. In the remaining two-handed kickback injuries, most lacerations involved the left hand, the face, and the neck; yet proper hand grip and body positions will both prevent uncontrolled kicks and keep body parts out of the arc of the kicking saw.

In later chapters, we'll be addressing other causes of injury noted on the accompanying graph. For now, though, here are some common errors that serve as contributing causes of kickback, itself, while collaborating with one another to play a role in most forms of chainsaw grief:

• Using a chainsaw to cut brush or unstable portions of limbs
• Gripping saw handles improperly
• Cutting overhead
• Swinging the sawbar away from the last cut before the coasting chain has stopped
• Letting helpers work near the saw
• Forcing a pinched bar from a kerf, with engine revved
• Cutting from a ladder or in a tree
• Cutting near brush, limbs, rocks, metal, or power lines

Of course, there are many other errors to avoid. Yet if we were all to avoid those above, an-

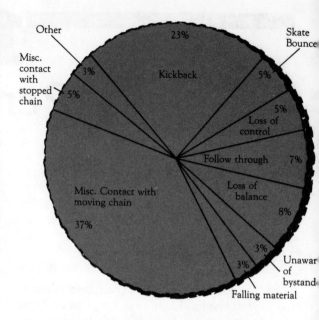

This pie chart, by the Consumer Product Safety Commission, shows causes of injuries determined through interviews with over 100 victims. Nearly all injuries involved chain cuts and all resulted in treatment at hospitals. Some categories deserve explanation. Skate/Bounce *means that the chain skated or bounced on the wood surface, usually on unstable wood, without making the intended kerf.* Follow Through *means that, as the chain completed the cut, it broke through out of control.* Miscellaneous Contact with Moving Chain *includes a host of causes: those involving coasting chain and those the victim wasn't really sure of; in this category, it's probable that kickback played at least a small role.* Miscellaneous Contact with Stopped Chain *includes cuts during sharpening, falls onto chain, etc.*

nual chainsaw carnage would likely shrink to a small fraction of what it has been.

In upcoming chapters, we'll be covering safe and efficient methods. Along the way, we'll try to note physical hazards, including those that often go unnoticed unless pointed out. Many of these hidden hazards involve stresses in wood, but there is an insidious hidden stress aside from wood. It often becomes a major hazard.

The Insidious One. Safety precautions seem reasonable enough when you read them or when a chainsaw expert explains them to you. If you are safety-conscious you will observe those precautions whenever you *begin* bucking in the backyard or working in timber. The insidious hazard usually grows in relation to the time you spend cutting on a given day. It grows in relation to the time your ears are subject to the muffled or unmuffled racket of intake, exhaust, and chain cutters on wood. In relation to the time your nostrils and lungs are subjected to exhausted fumes. In relation to the bending and lifting you do. In relation—*especially in relation*—to the ceaseless vibration your hands and arms undergo, transmitting this vibration to a lesser degree to your whole nervous system.

In a word, the most insidious hazard is fatigue. This is not the simple fatigue of physical work. It is unique to chainsawing. It is often more mental than physical. It dulls your perceptions as effectively as a martini or two. It leads to carelessness and even recklessness. Having experienced this unique fatigue, many sawyers end the workday without giving it another thought. Others may give it some thought only after the doctor has made the sutures.

The best way to combat chainsaw fatigue is to take frequent breaks or simply limit the amount of sawing you do each day. The breaks will help pump fresh air into your lungs and allow the tingling in your hands to assuage. The quiet, with the saw shut off, will let you plan your next series of cuts mindful of safety.

Hazards of bucking and limbing are greatly reduced after you clear away limbs, giving you and your saw more room in which to operate. Yet in a bucking situation such as this, high spots along the ground can cause powerful stresses inside a log. These stresses deserve careful analysis.

ASSORTED SAFETY AIDS

Owner's Manuals. Most owner's manuals provide some instructions for safe and efficient use of chainsaws. Normally, they will also explain how to get maximum protection from the particular saw's mechanical safety features such as the bar-nose guard, chain brake, reduced-kickback chain, and handguards. Of course, you're wise to heed the manufacturer's instructions as far as they go. The upcoming chapters reinforce the advice in most owner's manuals, while expanding upon it considerably.

Chain Sharpness. A sharp chain cuts efficiently and reduces wear on the guide bar, chain drive sprocket, clutch, and engine. A sharp chain also reduces the chance of kickback when the bar nose strikes wood because a sharp chain tends to slice through wood rather than bang and bash through. Since sharp chain cuts smoother, it also reduces the amount of fatique-inducing vibration you'll be subjected to, while reducing the time per cut in which you'll feel those vibrations. All things considered, the time you spend sharpening and touching-up your chain will pay safety dividends.

Saw Maintenance. If you are conscientious with maintenance tasks, you'll get best cutting efficiency and safety in the long run. Maintenance checks will help you ensure that your saw starts and idles well and delivers optimum thrust to the chain sprocket. With the saw running well, you'll be able to concentrate on your cutting and not be fretting about the saw's next coughs and sputters.

Conscientious maintenance checks also force you to examine closely the various parts of your saw. This will help you notice problems such as loose screws, a worn chain sprocket, and damaged bar-mounting threads. Such obvious defects are easy to correct, but if they go unnoticed they can result in saw damage and personal injury.

2
Specific Hazards and Safety Measures

A chainsaw is probably the most dangerous tool you can own. First come the hazards posed by the saw itself. Then come the hazardous situations you'll work in. Unless you pause to assess the hazards in each situation, you might operate your chainsaw mindful of only a few of them. Many of the hazards listed below are also noted elsewhere in this book. Here they are in a bunch:

Chain Cuts. These cuts include lacerations, amputations, avulsions, and decapitations. Cuts from powered chains usually occur lightning fast, leaving half-inch wide trenches contaminated by chain oil and wood chips. Cuts from coasting chain, after you release the throttle trigger, can be severe too. And the coasting chain itself can be kicked into you if it should accidentally contact resistance. An unmoving chain can cut you too, should you stumble and fall onto it or let your fingers nick cutters while you are sharpening them.

Lifting Injuries. Pulled muscles, slipped discs, and hernias undoubtedly increase with the woodcutter's age and zeal. Injury can occur if you attempt light tasks incorrectly—pulling a starter cord, hoisting a saw, tossing a stove-length log. You can also injure yourself tackling heavy loads, using poor form, or overestimating your own strength. Often, you can avoid injury by employing mechanical lifters or by first cutting logs to manageable sizes.

Overexertion. If you are unaccustomed to the rigors of woodcutting, take things slow and easy at first. Do some stretching and bending exercises. And be sure you've warmed up before attempting heavy work. Remember, physical exhaustion can lead to heart strain. And excessive perspiration can lead to chills and respiratory ailments. Besides, patient, methodical work habits can prevent overexertion, while also helping to prevent accidents that could result from fatigue and haste.

Losing Ground Footing. Lost footing is of course most dangerous with the chain at full throttle or else coasting. Yet many nasty cuts, as well as muscle pulls, occur during falls with an idling or shut-off saw. To maintain good footing you need foot gear with good traction and an ability to assess correctly how much traction foot gear can provide. When carrying the saw between cuts, anticipate what you would do with it if you were to slip. The safest measure is always to shut off the saw and slide the safety scabbard over the saw bar.

Falling from a Tree. This is essentially a residential-type hazard, greatly increased by ladders and chainsaws. The best advice is to stay out of trees if possible and, if not, to use only handsaws there.

Injuries from Falling Limbs. A dead limb shaken loose by saw vibrations or impacts from maul or wedges can crunch your skull like crackers. Called *widowmakers,* these dead limbs are aptly named. Professional loggers often pass up trees with many high dead limbs, and always wear hardhats when working under any tree, dead limbs or not.

Eye Injury. Flying sawdust can irritate eyes and lead to unsafe saw operation or eye injury. Flying chips can scratch eye tissue. And flying metal from a maul or ax on wedges can become embedded in the eye. Safety options include protective glasses, goggles, and face screens.

Abrasions and Splinters. Log wrestling often results in minor abrasions to unprotected hands and wrists. "Minor" puncture wounds and imbedded splinters present danger of infections. Best bet is to remove splinters promptly, and clean and dress all wounds. Sturdy work gloves and long sleeved shirts are the most reliable preventives.

Electric Shocks. The ignition system of a gas saw can give you an eye-opening jolt all right. But the power source for an electric chainsaw can kill you. All electric outdoor tools should be double insulated, and the power cords should be plugged into a circuit protected by a ground-fault circuit interrupter (GFCI), which can cut off flow of current within milliseconds after it detects a leakage of more than .006 amperes. Such a leakage might be a flow to ground through any object not intended to carry it, including you.

Falling Log Ends. Stove-length log ends of only eight-inch diameter can fall with enough force to smash toenails, bones, and ligaments. As footgear, sneakers and sandals are out. Sturdy wook boots are in, preferably with steel toe caps.

Falling Trees. Trees notched to fall north sometimes fall south, or east, or west. Even if a tree begins to fall in the designated direction, and even if bystanders are a safe distance away, a tree also can split upward at the cuts and jump back toward the faller. Or the asymmetry of its limbs or trees it hits while falling can cause it to twist and jump wildly rearward or sideward. Or an insufficiently high step resulting from an improper backcut could let the trunk jump backwards off the stump. Yes, felling can present many hazards, even if the trunk isn't partially rotted, even if the wind isn't blowing, and even it you make your cuts precisely by the book.

Exhaust Fumes.   Gas-powered saws emit the same air pollutants as autos burning gasoline, though without anti-pollution devices. But, riding in an auto, you are sealed off from exhaust gases. Operating a chainsaw, you breathe the pollutants that the wind doesn't disperse. These pollutants deserve concern:

Carbon monoxide: Colorless and odorless, this poisonous gas combines with blood hemoglobin, limiting the amount of oxygen your blood can absorb, causing oxygen starvation. Initial symptoms or poisoning include apathy, fatigue, headache, loss of visual acuity, and decreased muscular coordination—*all factors that can lead to chainsaw accidents.* Prolonged exposures at low and moderate levels are believed to cause brain and heart damage. To avoid poisoning effects yourself, never work in an enclosed area. Outdoors, try to work upwind of your saw. And since carbon monoxide is denser than air, avoid working in breezeless hollows. In addition, take frequent fresh-air breaks and keep daily sawing sessions to a minimum. Saw maintenance can help too: Since carbon monoxide is a product of incomplete combustion, a well-tuned saw will emit less of the poison than a poorly tuned one.

Other serious pollutants: These may have less immediate effects on you than carbon monoxide, but the cumulative long-range effects can be pronounced. Sulfur oxides and airborne particles cause the choking effect from exhaust fumes and aggravate heart problems and a wide range of respiratory ailments. Nitrogen oxides cause structural damage and chemical changes in your lungs, and also react in sunlight with hydrocarbons to cause smog. Lead accumulates in bone and soft tissue, affecting blood-forming organs, kidneys, and the nervous system. To avoid chainsaw-related problems from any of

Carbon monoxide from your saw's exhaust can begin to affect you after only a short time. Since carbon monoxide is heavier than air, avoid working in breezeless hollows for extended periods. At other times, try to work upwind of the saw whenever safe and practical.

these air pollutants, use the same precautions noted above for carbon monoxide.

Explosions and Flare-ups. A tree falling onto a gas can could cause an enormous explosion. So could tobacco embers near the fumes of spilled fuel. As well, many people commonly test a spark plug by removing it and placing its electrode near bare metal on the engine, and then pulling the starter cord to check for a spark; this practice can cause a flare-up of spilled fuel or of fuel that is being expelled from the open cylinder. Besides, if an engine with electronic (capacitor discharge) ignition is cranked with the spark unable to jump to the engine, the capacitor can become overcharged, ruining the costly electronic module.

Sparks emitted from mufflers can ignite fumes from spilled fuel or from open containers. Thus, it's wise to refuel away from the cutting site and to keep the fuel container safely away from both cutting and refueling sites.

Hot-metal Skin Burns. Exhaust gases heat the muffler enough for it to give you a nasty burn. And friction from cutting can heat up a saw chain and bar enough to burn you too. As preventives, wear work gloves or allow hot parts to cool down before touching them barehanded.

Errant-spark Brush Fires. Mufflers will emit burning particles large enough to start brush fires that can grow to holocausts. For this reason, foresters sometimes ban the use of chainsaws without special spark-arrestor screens. In fact, when forests are dangerously dry, authorities may ban the use of chainsaws altogether.

Vibration Injury. Loggers call it white fingers or TVD. Physicians may refer to it as Raynaud's syndrome or traumatic vasospastic disease (TVD). The symptoms usually include pale, whitish fingers that have lost their normal color because vibration has caused arteries to contract. Additional symptoms usually include numbness and, later, pain. You may also have the sensation that your hands have been transformed into rubber gloves filled with hot water. Pain is usually most evident in cold temperatures, when reduced circulation becomes a problem anyway, aside from that induced by saw vibration. Veteran loggers, after prolonged exposures, may experience excruciating pain that may recur any time they attempt to operate a saw; thus, white fingers can become a permanent occupational disability.

If you're cutting only enough wood to heat your home each year, it's unlikely that vibration exposures will give you a case of white fingers. Susceptibility has been linked to factors such as relative vibration of the saw itself, work methods, and even heredity, smoking, and diet. If you are susceptible, you could experience mild symptoms after even an hour of continuous sawing. Best preventives are to take frequent breaks and to use only saws with antivibration cushions, which serve to isolate the powerhead from the saw handles. And remember, vibration can lead to mental and muscular fatigue, making you more vulnerable to accidents.

Poison Plants. Plants that cause skin irritations and water blisters occur throughout North America. The big three—poison ivy, poison oak, and poison sumac—all contain a substance called urushiol that causes the poisoning. Some woods workers suffer severe cases of dermatitis from the slightest contact with clothing that has brushed against the plants. Other workers seem relatively immune, but tests have shown that repeated exposures can break down that immunity.

Best preventive measures include an ability to identify, and to avoid contacting, poisonous plants in your region. Also, gloves and clothing can prevent contact. But if you suspect that clothing has become contaminated, remove the clothing without touching the exterior and then wash it with several changes of water and strong soap. Do not wash it with uncontaminated clothing, especially underwear.

You can also apply protective lotions that help prevent direct skin contact and that can be easily washed off. Special neutralizing soaps are available and should be used immediately after you finish work. Extracts from poison plants are also said to have immunizing effects, whether taken orally or by injection. Before trying the extracts though, get the okay from your physician.

If you are susceptible to plant poisoning, be

especially wary of woods work in spring and summer. As well, be careful with logs delivered to your home. Although they may have been stripped of poisonous vines, they may still bear traces of the irritant.

Insect Stings. In the U.S., more people die each year from insect stings than from snakebite. Many of the victims are part of the less than one percent of the population abnormally allergic to insect stings.

If you have had alarming reaction to stings (more than a lump and an itch), consult your physician to determine if you have developed a permanent allergy and, if so, to determine how many of the five stinging insects your are allergic to: yellow jackets, honey and bumble bees, white- and yellow-faced hornets. If you are allergic, your physician can prescribe a series of injections of diluted venom that will desensitize you. You may then have to stay on maintenance doses indefinitely.

Otherwise, your physician may prescribe an insect sting kit containing a syringe preloaded with epinephrine (adrenaline), as well as instructions, a constricting band, and antihistamine tablets to help minimize reactions. Such a kit is an essential first aid item for woods crews or any group traveling far from medical facilities.

To prevent being stung, be wary of all bee, wasp, and hornet nests. Hanging papery nests and mud-daubed nests are easy to spot. Underground nests are not and may be betrayed only by the flight of insects in and out. So watch your step. Honey bees that have escaped man-made hives nest in hollow trees, so you can avoid such snag trees for safety's sake as well as in deference to the vital role that snag trees play in the whole chain of life.

You can also make yourself less attractive to stinging insects if you avoid wearing brightly colored clothing and sweet fragrances.

If you are stung, scrape the stinger out, rather than pull it out, because the squeezing of the stinger while it is being pulled will force more venom into the wound. Then wash with soap and water before applying cold packs and baking soda. Seek medical help if abnormal symptoms occur such as severe swelling, widespread hives, itching about the face, difficulty in breathing, headache, nausea, or dizziness.

Ants, flies, and mosquitos administer stings and bites to which some people are allergic. Aside from avoiding woods work when the critters are in top form, you should wear insect repellents and tight-fitting pants cuffs and shirt sleeves. Flies and mosquitos are attracted to bright colors. Black flies are so strongly attracted to dark blue that it's almost suicidal to wear dark denim in black-fly season.

Spider Bites. All spiders inject venom, but only two spiders present any threat to woodcutters. These are the brown recluse spider and the female black widow. To avoid bites, avoid reaching into dark places such as wood hollows or piles of wood or rock. If bitten by any spider, try to kill it for identification. Ice packs will slow absorption of venom. But any bite should be treated by medical people because even mildly venomous bites could cause infection.

Ticks. Woods workers in nearly all states face tick problems in late spring through summer in dry climates—into fall in humid climates. These leathery little arachnids look replusive enough. Furthermore, their bites can lead to infection and various fevers. Although only one to five percent of ticks carry the microbe *Rickettsia rickettsi,* known to cause the dread Rocky Mountain spotted fever (RMsf).

Of the over 1,000 cases of RMsf each year, over 90 percent occur in the eastern states. These result from bites of the American dog tick. A western species, the Rocky Mountain wood tick, may cause up to 10 percent of RMsf cases each year as well as lesser fevers. The southern lone star tick is suspected of causing RMsf cases as well.

As with humans, so with ticks, the female of the species is the woods workers only worry. Soon after mating, the blind and deaf female tick climbs to the tip of a twig or weed and, with forelegs outstretched, begins her trance-like vigil. She may wait for days, months, and even years until a butric-acid odor—common to all mammals—

arouses her. Thus aroused, she latches onto any mammal brushing close enough.

If you're the mammal, the blood-thirsty lady begins her quest over boots, chaps, and shirt. Any opening will do. That failing, she may reach your head. She searches for a tender spot and then releases a mild anesthetic as she bores unnoticed with her mouth's recurved barbs. She'll drink her fill in a day or two, nourishing the thousands of fertilized eggs within her. Engorged almost to almond size, she drops off, lays her eggs, and dies. If she carries fever microbes, so will her offspring. The longer she remains embedded, the greater the likelihood that she'll transmit the disease to you.

Techniques for removal of embedded ticks vary. The Red Cross and other authorities suggest that you force the lady to extract herself by smearing her with a heavy oil (mineral, cooking, machine) preventing her body pores from breathing. More anxious for air than blood, she's supposed to pull out. If she fails to respond, you're advised to wait half an hour and then attempt to gently pull her out, gripping her close to the head with tweezers. Some people burn lady tick's arse with a cigarette in an effort to make her move. Others kill her outright with tick repellent, gasoline or kerosene, but these methods are more likely to leave the tick's head embedded after your pull. *Note: Try to remove the head and mouth parts intact.* Otherwise, they'll likely fester.

Symptoms of Rocky Mountain spotted fever appear 3 to 14 days after the bite. They include chills, a run-down feeling, and a measles-like rash beginning at the extremities. A sore black scab may cover the bite, and red streaks may radiate from it. If some of these symptoms appear, you still might not have RMsf but may have contracted one of the less severe tick fevers. Fortunately, if administered in time, antibiotics are usually effective in killing the microbes.

To avoid ticks, some people simply stay out of tick territory in season. There are also tick vaccines. But to ward ticks off, you can tuck pantlegs inside boots, button shirt sleeves and collars, and apply tick repellent near openings in clothing and to your hair and beard. Home again, have someone closely inspect your head. Then strip, check your clothing, and inspect the rest of you, using a mirror where needed.

After removing an embedded tick, burn it and wash your hands thoroughly. Wash the wound, and then ask your physician if he or she wants to see you.

Snakebite. Less than 20 percent of the North American snake species are poisonous. Of these nearly all are pit vipers (rattlers, copperheads, water moccasins) whose venom is a hemotoxin (attacking the blood). The only other poisonous viper is the brightly banded coral snake of the South, which is related to cobras and carries a venom that is a neurotoxin (attacking the nervous system).

As to snakebite, woodcutters are pretty well off because the noise of the work usually scares snakes away. Also woodcutters usually work near vehicles that can speed them to medical help. Since many nonpoisonous snakes can be mistaken for poisonous ones, best advice is to avoid all snakes that you can't positively identify as non-poisonous. Otherwise, watch your step in the woods, and wear sturdy trousers and boots that protect the calf and ankle somewhat. In particularly dangerous snake country, you might want to wear snake-proof ballistic nylon leggings that do double service—both against snakebite and chain cuts.

If you or a partner is bitten, stay calm. Immobilize the bitten part, trying to keep it lower than the heart. If the bite is to a limb, also apply a constricting band (not a tourniquet) a few inches above the bite, only so tight as to allow you to slip a finger underneath. Apply a suction cup if you have one and transport the victim to medical help. If medical help is more than a few hours distant and if the bite causes severe pain, much swelling, shortness of breath, or shock, make one shallow incision through each fang mark parallel to the length of the limb and apply suction. If necessary, treat the victim for shock and apply mouth-to-mouth resuscitation.

3
Safety Gear

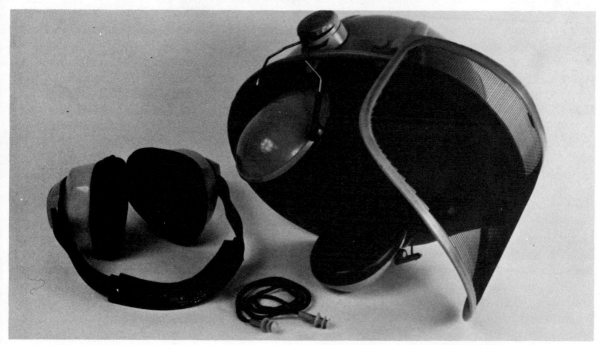

FOR THE HEAD

Chainsaw dealers and mail-order suppliers offer a full range of options that you can consider in relation to the kinds of woodcutting you do. Mail-order suppliers are listed on page 144.

Eyes. Chain rotation tends to throw wood chips and sawdust down, away from your eyes. But saw position and wind during felling cuts can send sawdust into your eyes. Of course, eye irritation caused by sawdust is a nuisance, but it can also be a distinct hazard because it may require that you either continue a cut with impeded vision or that you abort a cut and tend to the irritation, perhaps aborting a backcut on a strongly leaning tree. Aside from sawdust, flying metal from maul-

Principal choices in ear protection include muffs, plugs, and hardhat muff attachments. For most people, muffs are more effective than plugs for deadening high-decibel noise. In cold temperatures, plastic headstraps on muffs may become so brittle that they crack when flexed much. The string connecting the pair of ear plugs shown lets you drape the plugs conveniently around your neck. Otherwise you'll often be bothered with placing plugs into a dirt-proof canister and then retrieving them.

and-wedge work presents the other hazard to eyes.

Normal eyeglasses or lightly tinted sunglasses do pretty well in warding off sawdust and wood chips. Yet for any metal-to-metal impact work, safety glass or goggles are essential. There are also metal-mesh and nylon-mesh face screens that can be hinge-mounted to hardhats. These screens provide good wood-chip and sawdust protection, while allowing good visibility and ventilation. Face screens are dandy until a mosquito or bee flies inside.

Ears. Decibel counts at ear level of chainsaw users normally range between 90 and 120. Highest-decibel noise usually comes from exhaust ports, but noises from sprocket on chain, and chain on wood, often exceed 90 decibels too. Since prolonged exposure to noises exceeding 85 decibels can cause permanent loss of hearing, you're well advised to use noise deadeners whenever you use a chainsaw—even an electric chainsaw.

Well-fitting ear plugs can reduce noise to safe levels. Silicone rubber is more comfortable than plastic and forms a better noise seal. The more expensive plugs have valve openings that hold out high-decibel noise while admitting lower-decibel sounds—such as conversation and warning shouts. The valve also reduces the possibility of dizziness that sealed plugs can cause, and eliminates problems that could result from internal/external temperature differences.

The chief drawback of ear plugs is poor fit in one ear or both. Better plugs come in small, medium and large diameters—though mediums fit most people best. Some plugs have a neck cord, attaching the plugs, that runs around the back of your neck. This helps prevent loss and allows you to remove the plugs when not sawing, leaving them draped conveniently around your neck. To ensure cleanliness and to help prevent loss, most plugs makers supply small carrying cases.

Ear-muff protectors may cost twice as much as the best plugs, but they offer a surer noise-muffling fit for most people. Of course, muffs are more cumbersome and, in summer, hotter. You can buy the adjustable-headband type with flex-position muffs or the type that mount on hardhats, which are best bought mounted. (If you test various plugs against a muff, you are likely to notice superior noise reduction with the muff.)

Here's a further way to gain appreciation for plugs and muffs. Operate your saw for a short while with no protection. Your nervous system will probably adjust to the noise sufficiently for you to think the noise is bearable. Next wear ear protection for a goodly cutting session and then, for an instant, remove the protection. You will be surprised to note that the noise, even at idle, now seems unbearable. With that proof, you'll probably insist on using protection ever after.

Skull. Hardhats are essential for most felling work and for any groundman. They are available in wide brim or narrow brim. The narrow-brim hat has become increasingly popular with wood-cutters because it is better suited for attachment for ear muffs and face screen. Hats come in aluminum alloy and high-impact plastic, each type with adjustable headband.

Some woodcutters economize initially by buying either a hardhat or ear muffs or goggles, figuring on adding to the safety arsenal in time. But muffs with adjustable headbands do not allow you to wear a hardhat on top. And if you buy only a hardhat, you'll be stuck with ear plugs or the eventual task of installing muffs on the hat. If you plan to go first-class, in time anyway, consider buying a muffed and screened hardhat right away. Besides, you'll probably find other uses that justify the cost, even if that means snoozing under a tall hickory when the nuts are dropping.

FOR HANDS AND ARMS

Work gloves usually serve better than mitts, except in cold temperatures when you need mitts to reduce loss of heat from your fingers. Work gloves can help in many ways. They protect hands from punctures, splinters, and muffler burns. They also protect untoughened hands from blisters and abrasions. Gloves add a small, though valuable, cushion against vibration, and they keep your hands relatively free of oil smears. In chilly temperatures, gloves help your hands maintain warmth and circulation; otherwise cold and vibration can team up to contract arteries, inducing symptoms of the white-fingers disease discussed earlier. Gloves also absorb sweat and allow a better handle grip.

These "chainsawing" gloves were mounted over the padded end of a hockey stick before being dropped onto a fully throttled chisel chain. The glove at left had only one layer of nylon and offered no protection, as you can see. The mitt at right was dropped onto the revved chain twice. It had four layers of ballistic nylon, which prevented chain penetration through the final layer in one test and allowed only slight penetration in another.

As to materials, avoid cotton palms because they afford poor grip, abrade fast, and become clammy from perspiration. Inexpensive gloves with roughout leather palms and finger tips, at least, and with cotton backs and cuffs serve pretty well for average firewood cutting. Here, avoid super-cheap gloves with thin palms and loose stitchery because these might not last a few hours. Logging-supply catalogs advertise inexpensive timberfaller's gloves made of wrist-length knitted nylon that are washable and reversible. But they abrade easily and should only be used as liners for leather gloves.

In gloves and mitts, you will usually get what you pay for. Top-quality woodcutter gloves and mitts have double-leather palms, ballistic fabric layered backs, and reinforced stitching. Velcro fasteners let you tighten cuffs to keep sawdust and wood chips out. Such quality handgear can cost as much as a saw chain, though.

For your arms, even in hot summer, wear long sleeve shirts. This will keep wood chips and sawdust from sticking to perspiration, and it will help protect your wrists from abrasion and poison plants when you wrestle logs.

FOR THE LEGS

Depending on the studies you read, cuts to legs and ankles account for 33 to 48 percent of chainsaw injuries. In this area, ballistic leggings have helped some logging companies nearly eliminate bad cuts to the legs.

Informal tests conducted for the accompanying series of photos led to a sobering conclusion. Ballistic fabrics are definitely *not* chainsaw-proof. As of this writing there was no ballistic fabric on the market that would stop the chain of a throttled saw. Yet a large percentage of chainsaw cuts result from coasting chain during kickback and when a coasting chain is inadvertently allowed to brush the leg. Ballistic pads can offer significant protection. The reason for the pad effectiveness lies in its layering. The layers slip, slide, and stretch upon contact with chain, *sometimes* stopping a coasting chain, and buying you small fractions of a second before a throttled chain cuts through.

Ballistic pants are somewhat cumbersome to wear, and in warm weather they may seem too hot. Thus, many woodcutters prefer ballistic chaps that promote air circulation, but chaps have straps and shell edges that can hang up in brush. Beware of advertised *lightweight* summer pads that may offer only a couple of layers of ballistic fabric. Remember, that ballistic fabric is effective

Ballistic leg pads are available as trouser inserts, like these, or as chaps. The inserts are removable should you want to wash the trousers. But if you wash inserts, they require a long time to dry.

For this test, Daniel Tilton nailed a roll of floor carpeting to a log on a sawbuck. A four-layer ballistic-nylon trouser insert was then taped to the carpeting. Tests included (1) full-throttle drop of the saw bar onto the padding, (2) a drop of a full-speed coasting chain, (3) a drop of a chain that had been allowed to coast a couple of seconds. Invariably, the throttled chain cut through all padding and into the carpeting. The unthrottled full-speed chain cut barely through or almost through on most tests. The chain that had been allowed to coast a couple of seconds usually penetrated only about half of the ballistic layers. How does four-layer ballistic nylon compare to other materials? Drop tests Tilton conducted on wood and rolled carpeting seem to indicate that a four-layer ballistic pad could offer about as much protection as a one-inch slab of pine or three layers of carpeting. That's not invulnerability to be sure, but it's a significant advantage over denim. (After the above tests, independent testing labs published reports ranking ballistic Kevlar higher than nylon; lighter weight, better protection.)

in relation to the amount of slipping and sliding the layers do before letting the chain penetrate.

FOR THE FEET

As a minimum, always wear a sturdy boot or high-topped work shoe with good traction. Steel toe caps can help ward off chainsaw cuts, as well as dropped log ends and errant maul and ax blades. Special logging boots also feature a lining of

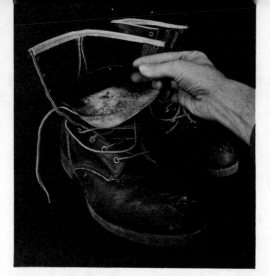

Special logger's boots available from forestry supply catalogs feature stainless steel toe caps and four layers of ballistic nylon in front—the padding shaped like a broad auxiliary boot tongue. The protective materials add weight and the nylon results in a bulky, stiff feel around the ankle. But such boots would prevent serious injury from impacts with coasting chain.

ballistic fabric. For climbing, stiff soles with steel support plates protect feet from the pressures of ladder rungs and tree crotches.

CLOTHING INCIDENTALS

Some of the worst hunters and fisherman are the most magnificently clad. Perhaps fearing this stigma, some woodcutters are reluctant to wear the clothing and gear discussed above. Yet none of that above is merely stylish. It's *safe-ish* though. Here the dollars spent are not frivolous and won't make you look frivolous unless you also shop for a whole new wardrobe in which to perform your cutting chores.

Old, serviceable pants and shirts do fine, provided they're reasonably close-fitting and have no rips and flaps that can become hung up in branches or saw chains. Many pros prefer to wear suspenders and loose-waisted trousers that allow freedom of movement and room to tuck in layers of wool shirts. Wool doesn't stand up to abrasion well, yet a layer of wool next to the skin will wick sweat away and will retain much of its insulating value even when damp. Wool and polyester blends are tougher.

4
Proper Cutting Form

Your cutting form can greatly determine your cutting efficiency, degree of fatigue, and personal safety. With this in mind, consider these three guidelines:

1. Always maintain a solid two-hand wraparound grip while the chain is moving. Otherwise, if the unprotected nose of the guide bar hits resistance, even an unthrottled coasting chain can cause an injury-producing kickback.

2. Keep your left hand on the left handle and your right hand on the right handle for all types of

Roy Booth Sr., a champion at Homelite's annual Tournament of Kings, demonstrates flawless bucking form: feet set for best balance, left thumb wrapped under the handle, left elbow locked, back saw handle resting on thigh as convenient, all body parts aside from the plane of the saw bar. (Note: In contests such as this for big money, competitors usually spurn safety gear even though they wear safety gear in woods work. For example, ear protection would muffle the starting signals. And gear for legs, hands, and head could cost precious fractions of a second in close heats.)

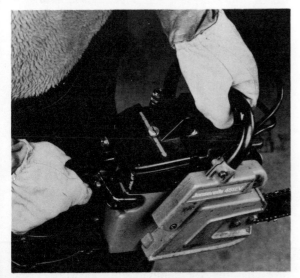

For all cuts, grip the saw firmly with two hands—thumbs wrapped under the handles.

cuts. This helps ensure that you avoid a cross-over grip that can put your arms, legs or upper body dangerously into the plane of the rotating chain.

3. When using a saw without an antikick bar-nose guard, be aware of the path of the bar nose as well as the progress of the cut.

Vertical Cuts. These include overbucks on horizontal-lying logs using the bottom edge of the chain bar and underbucks using the top edge.

Overbucks are simplest when you can make them straight through a log, topside to bottomside. You can make straight-through overbucks whenever a log is held firmly in place at a support either to the right or left of the saw. As you complete the overbuck, the unsupported end of the log falls to the ground.

Yet, you may often have to make *partial* overbucks in deference to saw clearances or log stresses. When there is insufficient saw clearance to overbuck straight through, you may decide to overbuck partway in several places along the log's length before rolling it to overbuck the uncut portions. *Hint:* If you elevate a log before rolling it, you'll keep the bark from picking up chain-dulling grit.

Internal log stresses are determined by the points of support under the log. For example, when a log is supported far to the right and to the left of your intended cut, there will tend to be a sagging sort of stress at the cut. The longer the log span in relation to log diameter, the more effect the stress will have. In this case, if you attempt to overbuck straight through, the sagging stress might cause the top edge of the kerf to close, pinching the saw bar. This pinch can be as unyielding as a vise—stopping chain rotation and preventing removal of the bar. To relieve the pinch, you may either have to drive a wedge into the kerf or raise the log— thereby opening the kerf—or cut the bar free using another saw.

It's better to avoid a log's sagging stress if you can by supporting the log near the intended overbuck.

An alternative is to overbuck a short distance and then underbuck toward the kerf made by the overbuck. As the combined cuts weaken the log, it

Daniel Tilton, instructor in Swedish logging techniques, begins the first of two face cuts to form a felling notch. Note how he braces himself against the tree and supports the back handle with his thigh.

will gradually sag, conveniently opening the kerf for your underbuck. Beware though of the falling force of the log ends on your chain bar. More on this factor in the chapter on limbing and bucking.

Felling Cuts. These perhaps offer more potential for mistakes and disaster than vertical cuts. Here you have to consider how the tree will fall, how to make a precise series of cuts, how to hold the saw safely, and how to brace yourself for best saw control while still being able to see your work.

Again, remember the three safety guidelines:

Ron Johnson here bores into the first of three balloons he must break in one of six events in Homelite's Tournamet of Kings. In woods work, bore cuts are often used to start cuts from the middle of a log toward an outer edge; this is done on felled trees to relieve stresses caused by points of support under the log. Consider Johnson's ideal stance: feet spread, left elbow locked, thumb under left handle, head aside from the plane of the saw bar. (Note: Outside of competition, champions such as Johnson recommend the use of safety gear.)

Tilton here uses the topside of the bar to make the backcut on a small red maple. In this position, the chain exerts a noticeable push toward Tilton, so he braces his right wrist against his right knee.

Bore cuts must be started with the underside of the bar nose, as shown. According to world-class lumberjack Tom Bildeaux, "If you run the bar into the wood at the correct angle, the chain will draw itself in like a car with snow chains on." But if you nick the log with the front or top of the bar nose, the bar will kick upward at you. If you gain control over the kick before being cut, you may still receive a jolt from the handles that travels through your whole body. Aren't you glad you kept your left thumb under the left handle and your left elbow locked?

(1) two-hand grip, (2) no cross-over grips, (3) readiness for kickback.

Next, prepare to make your felling cuts. If you are cutting hardwoods for firewood, the lowest ten inches of bark will often be gritty from rain splash. Use a hatchet to trim this bark or rub it with a steel brush. Otherwise, the grit may quickly dull your chain, forcing you to resharpen before proceeding with the cuts. There is usually no grit on low-lying bark of conifers because the evergreen canopy prevents rain splash there.

For more quidelines and illustrations on felling, turn to the chapter on felling.

After having made a face notch (hidden by the saw's chain brake here), Tilton begins a backcut by boring straight inward to form a hinge (later described in detail in the chapter on felling). He will next cut rightward toward the tree crotch. In this case, the crotch allowed Tilton no room to make a conventional backcut. Note Tilton's hand and body positions, the right thigh supporting the right hand. Similar backcuts are made on strongly leaning trees in order to form the hingewood. Then the wood on the opposite side of the lean is the last to be cut. This helps prevent the tree from going too soon, possibly splitting upward in the dread phenomenon known as barberchair.

When tree diameters are more than twice the length of the saw bar, professionals may find it necessary to bore into the holding wood in the tree's core before starting the backcut. Note hand and body positions, especially the thigh support of Tilton's right forearm. From this, you can see why nonprofessionals find it easier and safer to make felling cuts with a guide bar at least as long as the tree diameter.

Boring Cuts. These are drilling or stablike cuts. They can be relatively difficult and dangerous because to start them you must have precise control of the potentially unruly nose of the bar. Many safety experts advise that nonprofessionals avoid boring altogether. Yet boring skills offer the sawyer greater versatility, especially when making felling cuts in large trunks and in relieving stress and tension in felled trees. Best bet is to practice and master boring techniques for felling on stumps before you attempt them on standing trees.

Before you begin the cut itself, brace yourself, totally ready for a vicious kickback should the upper portion of the bar nose nick any wood. Begin cutting with the bottom quadrant of the nose. Plan and manipulate the cutting so that the nose will soon be buried in the wood. Then if the saw should kick, the cut itself will serve as a sheath, allowing only a linear kick (a push really) if the chain along the upper quadrant of the bar temporarily hangs up. To avoid control problems, never bore above waist level.

Note: With some reduced-kick chain, boring may be tedious if not impossible. This is because the hinged guard link in front of the depth guage swings outward as the chain arcs around the nose, thus preventing the cutters from taking good bites.

Kickback Savvy

This simulation of the classic rotational kickback was achieved with multiple exposures of camera film. Skilled sawyers can intentionally cause the initial thrust of such a kick, minimizing the rotational arc of the saw by absorbing forces through their hands and arms. Unprepared for such a kick, you could lose control of the saw, perhaps sacrificing your left hand or arm to the voracious chain. Many kickback cuts also occur to shoulder, neck, and head. Severely kicked saws have even whipped around far enough to put trenches in sawyers' shoulder blades. Modern mechanical safety features on saws can greatly reduce the incidence and force of saw kicks. But the surest preventive is your own skill and vigilance.

CAUSES AND FORCES

Hospital reports on chainsaw injuries may quote the victim as saying the saw jumped up, or flew up, or bounced back, or jerked up. Such descriptions are traumatized laymen's equivalents of the troublesome phenomenon the chainsaw industry calls *kickback.*

Kickback occurs whenever the parts of the rotating chain momentarily hang up on an obstruction. Technically, kickback can occur anywhere along the exposed chain's path.

Classic rotational kickbacks occur when the chain parts meet resistance as they pass around the upper quadrant of the bar nose. Here cutters and depth gauges are arcing forward at unsuitable cutting angles, the space between cutters larger than desirable. A momentary chain hangup ricochets engine torque to the saw. And this often causes an abrupt and violent upward arcing thrust of the saw bar.

Less violent kicks (pushes and pulls, really) can occur anytime the cutters along the flats of the bar meet obstructions. This may occur when the chain hits the bottom of a previous cut or the side of the saw kerf, or when the kerf pinches closed onto the chain. When you are underbucking a log—using the topside of the bar—the saw tends to push toward you. When overbucking—using the bottomside of the bar—you'll feel a pull, which the saw's bumper spikes will halt. The firmer your grip and the more stable your stance, the less effect these pushes and pulls will have.

Besides kicking in one direction, a bar powerfully kicked could hit another object and surprise you then by ricocheting in the opposite direction—a double kick. Unnerving at best!

PREVENTIVE KNOW-HOW

Correct hand grip, body position, footing, and an anticipation of the kicking force will help you handle and absorb the force without letting it whip the bar. In addition, observe these precautions:

- Avoid letting the nose of the bar accidentally contact obstructions.
- Either swamp out (clear) the area of possible obstructions or carefully note them beforehand so that the chain cannot make inadvertent contact.
- Avoid using an excessively long saw bar for limbing and most bucking unless it is equipped with an antikick nose guard. An unshielded nose, moving about just beyond focus of your eyes, can blunder onto obstructions.
- Keep chain teeth sharp. Dull teeth tend to butt their way through wood, leading to excessive saw vibration and hangups that can cause kicks.
- Avoid cutting from awkward or unstable positions. Instead try to reposition either yourself or the log.
- Rev the saw to full power just as the chain is about to make first contact with the wood. A slow-moving chain tends to butt against wood and thus hang up and cause a kick.
- When starting a cut using the top of the bar (underbucking), be prepared for a backward and sometimes downward pushing force from the saw.
- Let a coasting chain come to a full stop before you remove it from the kerf and before you swing the saw bar to a new position. Never relax your two-hand, opposable-thumb grip until the chain has come to a full stop.
- If you feel fatigue setting in, quit sawing. Mental and physical fatigue can make you vulnerable to kickback forces and to mistakes that cause the kicks.

PREVENTIVE MECHANICAL DEVICES

With kickback having been largely responsible for over 30 percent of chainsaw-related injuries in some years, manufacturers have competed to offer features that would help hold kickback carnage down. Here are some of the most effective devices.

At present, the surest mechanical preventive of rotational kickback is a bar-nose guard that prevents chain from hitting resistance when arcing around the upper quadrant of the bar nose. The guard shown is known as the Safe-T-Tip and was introduced by Homelite.

Most saw manufacturers offer a chain brake assembly similar to this Husqvarna brake. Ideally during uncontrolled kickback, pressure from your left hand onto the handguard shown causes the steel band to close onto the clutch drum (shown on saw in background). This action is intended to stop chain rotation before the chain reaches you. There are also "automatic" inertial brakes that are designed to activate whenever sudden kickback forces are applied to the saw bar.

Antikick Bar-nose Guard. This is a fenderlike device mounted over the bar nose. On some shorter bars, the guard may come permanently mounted. On most bars it is detachable by means of a bolt. Mounted in place it eliminates rotational kickback because it prevents obstructions from hitting the chain traveling around the nose.

Although the nose guard eliminates kickback, it imposes significant operational disadvantages. First, it allows you to cut only through logs of somewhat smaller diameter than the bar length because the nose guard won't fit through the saw kerf. Second, the nose guard disallows boring (drilling) cuts. Thus, in order to make boring cuts or tackle logs of bar-length diameter or more, many sawyers remove the nose guard and forsake the antikick feature. Unfortunately, the inconvenience of removing and reattaching the nose guard for varied cutting chores leads some woodcutters to remove and then shelve the guard.

Yet if you are safety conscious and have a removable nose guard, you can plan work so that you tackle all small-diameter wood with the guard in place first and then remove the tip only long enough to accomplish big chores. If a bar nose guard were required on all saws, it would offer all sawyers a reliable antikick option on smaller logs.

Chain Brake. This device is designed to stop the chain under laboratory conditions within small fractions of a second after kickback begins. Typically, a chain brake consists of a left handguard linked to a lever and cam mechanism that can forcefully apply restriction of a steel band around the clutch drum, stopping the clutch and the chain.

Most chain brakes are designed to activate whenever a kickback is severe enough to cause your left hand to slip onto the handguard or your left wrist to bend forward onto it. Receiving only light forward pressure, the handguard will spring forward, applying the brake.

Some saw makers also couple an "automatic" inertial chain brake into the basic braking system. Here sudden kicking force applied to the bar actuates the braking mechanism, clamping the clutch drum to a halt and thus stopping the chain.

Although a chain brake has merits, it does not prevent kickback. Yet it can prevent and reduce the extent of kickback injury if it manages to stop the kicked chain before it hits you. Besides the braking function, the left handguard/brake lever can help protect your left hand should your hand slip forward or should the chain itself break apart, with one chain end whipping upward.

Critics of the chain brake contend that it does not usually stop the chain soon enough. They contend that to stop a chain before the saw makes an unimpeded 90-degree rotational kick, the brake must function within 10 to 20 milliseconds (10/100 to 20/100 of a second). If flesh and bone are within that 90-degree arc, they reduce the required braking time. Besides, precious fractions of a second are lost before the left hand slips forward and actuates the handguard lever. The oil and sludge inside the clutch housing, as well as incorrect brake adjustments and worn brake parts, could slow the brake or cause it to fail entirely. Inertial brakes, activated by a kick-like impact of the saw bar, can actuate the brake mechanism somewhat sooner.

Manufacturers of chain-brake equipped saws advise owners to clean and test the brake at least daily. They also advise that you employ proper hand grip and body positions for all cutting tasks.

If your saw is equipped with a chain brake, you have a worthwhile though not infallible safety feature. It does not prevent kickback, and it should not be regarded as surefire protection in the event of a kickback. It is merely one of several optional antikick saw features that can work in your favor should you make a big mistake, by carelessly allowing an unimpeded kick to occur.

Left Handguard. During kickback, a fixed left handguard is less effective than a handguard linked to a functioning chain brake. But a fixed handguard alone was credited with significant reductions in hand injuries among professional

A left handguard, as a minimum, *should be mandatory on all saws. This device helps guard the left hand during mild kickback and also if a broken chain end whips upward.*

loggers in Sweden in the early 1970s. In a mild kickback, in which your left hand slips off the handle, the handguard can catch your hand or wrist, preventing severe lacerations or amputations. In a violent kickback, a fixed left handguard could help somewhat too—perhaps only spraining your wrist if you are lucky. And, as with a chain brake, a fixed left handguard can also protect your hand from chain backlash should the chain itself break.

Reduced-kickback Bar Nose. Various bar and saw makers offer narrow-nose and nonsymmetrical, banana-shaped bar noses that reduce the arc of the chain passing over the dangerous upper quadrant of the bar nose. Tests have shown that with smaller noses, chances of kickback are reduced. But when a nose is simply narrowed, chain wear increases and boring speed decreases.

Makers of nonsymmetrical bars feel they've solved the problem by moving the smaller radius upward, giving you a desirably smaller nose radius in the dangerous upper quadrant, with no reduction in cutting performance in the lower quadrant. But this nonsymmetrical kickback-reducing nose presents drawbacks too. The bar should always be operated topside up and not flopped top-for-bottom after each chain sharpening as you are advised to do with symmetrical

The asymmetric (banana-nose) bar reduces the arc of the chain in the kickback-prone upper quadrant of the nose. Thus, it helps reduce probable incidence and severity of kicks—somewhat.

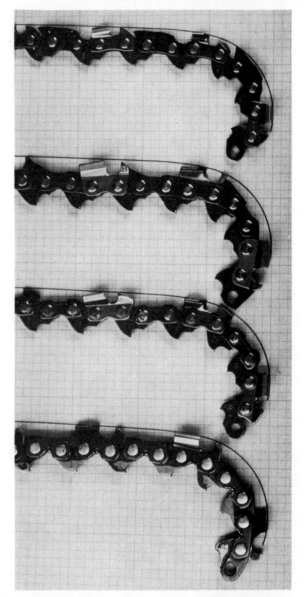

bars. The result is that the nonsymmetrical bar receives excessive wear on the bottomside because that's the side you use most. Then as a bar wears unevenly it also begins to wear chain excessively. Using such a bar could mean shorter bar and chain life, and in extreme situations operation with a dangerously worn bar and chain.

Occasional spokesmen for nonsymmetrical bars have indicated that you can mount the bars upside down without undue hazard, but such operation will reduce nose life and nose-buried cutting speeds. And it will also wipe away the antikick protection designed into the bar. *Note:* Replaceable nonsymmetrical noses can be flopped, top-for-bottom, on the original bar so that you can flop the bar to promote even wear. This is a job for a service shop.

Reduced-kick Chain. Chain makers now offer many chain configurations that can reduce kickback forces 59 to 74 percent over old standard chain. Here engineers have focused on three principal design factors: (1) cutter profile, (2) depth-gauge profile, and (3) tie-strap profile.

• *Cutters.* The higher the cutters in relation to other wood-deflecting chain features, the greater the cutters' tendency to hang up when they meet resistance. This is especially pronounced as the cutters arc around the bar nose. In this regard,

This photo shows profiles of old-standard chain (bottom) and a few of the reduced-kick chains as they arc around the kickback-prone upper quadrant of the bar nose. Tests have shown that sloping depth gauges protected by articulated ramps on non-cutting links can reduce kicking forces by more than 50 percent, while reducing kicking tendencies as well.

some makers offer low-profile chain, whose cutters are simply lower overall in relation to the bar rails. But lower-profile cutters arcing around the bar nose also present lower-profile cutters on the flats. This usually results in less voracious cutters and consequent slower cutting.

• *Depth Gauges.* Design innovations here have reduced potential kickback forces. A raked-back depth gauge provides a gentler-sloped ramp that better deflects wood prior to impact with cutter teeth. Such a depth gauge promotes smooth cutting around the bar nose.

• *Tie-strap Profiles.* One of the most effective kick-reduction features to be developed is so simple in principle that you might wonder why makers didn't introduce it long ago. Different chain makers offer variations. But basically, the tie-strap is a raked-back fin ahead of the depth gauge. As the chain begins arcing around the nose, the pivoting chain parts cause the tie-strap to swing outward, helping deflect wood, preventing jolting impacts with depth gauges and cutters. And this *reduces* the likelihood and severity of kickback.

SO WHICH ANTIKICK DEVICES ARE BEST?

The safest saws would be equipped with all of the devices described above. An antikick bar nose is great, *if it's mounted in place.* But serious sawyers will find need to remove it in order to tackle large logs and special boring chores. With no antikick bar-nose guard, the combination chain brake and left handguard is vital. Next assymetrical, banana-nose bars used with reduced-kick chain can add a margin of safety.

No matter what safety featurers your saw has, your skill and vigilance will be the surest kickback preventives. Remember, the mechanical antikick and reduced-kick devices are designed to help protect you less from a dangerous machine than from your own mistakes in using it.

6
Safe Refueling, Start-up, and After-the-cut Idling

REFUELING

Try to pace the various tasks of woodcutting to coincide with refueling. For example, when you are running low on gas or when you run out, set the saw down and take a break from the vibrations and fumes. This is a good time to touch-up saw teeth, pile wood, or just take a blow. (*Caution:* If you are a smoker, never smoke while using the saw or refueling it. Tobacco embers, of course, can ignite fuel fumes. Also, if you smoke while sawing, smoldering tobacco could fall unnoticed and start a fire amid sawdust or forest tinder.)

It's usually easiest to refuel a saw if you elevate it on a chopping block or flat-cut stump. This is especially important when working in snow or poorly cleared areas. Besides supporting the saw at a comfortable working height, the elevated platform lets you hold the fuel container low enough to prevent surging spills that often can result when the fuel is forced through the nozzle faster than you want it.

Before opening the fuel container, shake it up well to ensure a good mix of gas and oil. Brush your arms, gloves, and front free of sawdust and wood chips. Then remove your gloves, if temperature allows. Using a rag, wipe the fuel-cap areas on the fuel container and the saw free of grit and sawdust.

Remove the cap on the fuel container. Assemble the nozzle, and loosen the container's cap. *Note:* If the nozzle has a screen, check that it's clean. If the nozzle is designed to use a sponge filter, check that the sponge is inside the nozzle and clean; sometimes the shaking of the container and careless nozzle assembly result in the sponge's falling into the container, where it does no good.

When you have readied the fuel container, remove the fuel cap on the saw and set it down so that its thread and vent remain clean. Firmly gripping the container's handle with one hand, use your other hand to guide the nozzle into the saw's tank. Thus ready, slowly raise the container so that fuel begins to emerge gently from the nozzle. Avoid dumping fuel in, which can result in backwash and overflow. Watch the rising fuel level in the tank. As it approaches the top, gradually lower the container to reduce flow.

In case haste or inattention causes spilled fuel on the saw, be sure to wipe it dry. Then, first checking that the saw's fuel cap is clean, replace it. Finally disassemble the nozzle and replace it and caps.

Very Important: Remove the saw at least 15 feet from the cutting area. If any fuel spilled onto the ground or support block under the saw, do not attempt to start the saw there because sparks emitted from the muffler could ignite the fumes.

These combination 1½-gallon fuel and 1 quart oil containers are plastic, though some states condone only metal fuel containers. Vent caps prevent forceful surging of fuel during pouring. The nozzle assemblies on the unit at left disassemble for storage inside. This feature adds tasks and time to the refueling process, but it helps you keep nozzles clean and so helps prevent contaminants from entering the saw's fuel tank. The unit at right offers spigots, which are especially convenient in cold weather.

Fuel nozzles with sponge-filter inserts like this one reduce the chances that sawdust will enter the saw's fuel tank—provided you store the nozzle safely inside the fuel container and keep the nozzle clean when refueling.

Though of excellent design, this 2½-gallon fuel container is too large for nonprofessional use for several reasons: It holds about twice as much fuel as most nonprofessionals need in a year. In this case, the fuel would go stale long before use. Besides, partly empty fuel containers encourage water condensation, which contaminates the fuel. You'll be better off with a 1 to 1½-gallon container. It will force you to add fresh gas formulated for the temperature of the season.

Refueling aids: An old rag lets you wipe up fuel and oil spills, remove sawdust that could be blown into the fuel tank, and clean your hands of oil and sawdust. A rubber band cut from an old inner tube provides a holster for a scrench and a toothbrush—the brush here for cleaning a metal-mesh air filter.

This could easily cause a ground fire and even lead to an explosion of the fuel tank.

The above procedures may seem tedious, but they really add less than a minute to the refueling operation. Yet many woodcutters are careless when refueling and routinely risk catastrophe each time.

There's one more absolutely essential "refueling" task. What is it? You're right! It's also time to add oil to the chain oiler. If your saw's oil cap is on the topside of the saw, as the fuel cap often is, simply add oil using the same standards of cleanliness as you used for fuel. However, if your chain oiler cap is on one side of the saw, requiring that you lay the saw on its side to add the oil, always add chain oil before refueling the saw. Otherwise, fuel from a full tank may leak out the fuel-cap vent onto the saw. Spilled fuel—again—poses a flare-up and explosion hazard and so must be wiped off.

SAFE START-UP

Study your owner's manual for details on saw settings for start-up (ignition switch, choke, throttle interlock, chain brake). The manual should also provide instructions for stabilizing the saw while you pull the starter cord.

Remember, two common reasons for a saw's failure to start are not mechanical but human: (1) out of fuel and (2) ignition switch-off. Having ensured that these problems aren't at fault, you can determine other causes for starting difficulties by troubleshooting as explained in your owner's manual and later in this book.

Owner's manuals usually recommend two-point stabilization of the saw as you pull the cord. For small saws, it's usually best to place the saw on the ground, hold the saw in place by grasping the left-hand handlebar, and place one knee onto the rear of the saw for two-point stabilization. For larger saws and for small saws with right-hand handles extending from the bottom of the saw, you are usually advised to grasp the left-hand handle-

Before starting a saw on the ground, clear away tinder that muffler sparks could ignite. Stabilize the saw with left hand on left handle and right foot inside the right handle, as demonstrated here by Daniel Tilton, instructor in Scandinavian logging methods. Pull the cord straight upward through its guide, rather than abrading it at an angle. Also, avoid pulling the cord to its end, creating rapid cord wear there.

Carlton Chain Company's Brian Lepine demonstrates a safe stand-up method of starting—provided the saw has an extended back handle that you can clamp between your lower thighs. This method is often more practical than placing the saw on the ground. Caution: Do not confuse this method with the popular, but far more dangerous, drop-start method in which the saw is thrown down with the left hand from chest level as the right hand pulls the starter cord. The ill-advised drop start allows poor saw control in the event of an engine backfire and bar contact with a branch. And it is needlessly strenuous.

bar with your left hand and place your right foot inside the stirrup formed by the right-hand handle and handguard.

Other Tips. Because the muffler may emit sparks, scrape forest litter away to bare ground. Yet ensure that the chain cannot come into contact with rocks or brush as you raise up with the saw to begin work.

Be especially aware of the safe idle and half-throttle speeds. That is, you don't want the saw revving enough to engage the clutch, which will start moving the chain. If the chain moves at idle, your saw is unsafe to operate and needs adjustment of the carburetor, the throttle mechanism, or the clutch.

Keep the bar from dipping into dirt as you pull the cord. Grit from the ground will cling to oily cutters and dull them when you commence cutting.

AFTER-THE-CUT HAZARDS

Many injuries and near injuries occur because saw chains continue coasting at high speed for several seconds after the user releases the trigger. This coasting chain can of course cut flesh, and it can also be traveling fast enough to kick back if it strikes some other obstruction.

This after-the-cut hazard is especially noteworthy because as woodcutters gain confidence they begin habitually looking away from the revolving chain upon releasing the trigger. Here an accidental nick of the saw kerf or an unnoticed branch can bring trouble. Helpers too are often injured by coasting chains.

Remedy: Always maintain a solid two-hand opposable-thumb grip on your saw until you see the chain stop. Then you can swing the saw aside or reposition it for the next cut.

7

How to Make a Folding Sawbuck with Limb Vise

The buck shown is a special chainsawing version of a design dating back to the old bucksaw days. It's sturdy enough to handle any log you can lift. Yet it also clamps lightweight wood in place that would otherwise bounce and chatter under your chainsaw. Incidentally, the buck also folds flat for storage and serves well with all manner of handsaws.

The vise clamps down when you step on the foot lever, and it opens *automatically* when you remove your foot. In effect, the vise gives you an extra "hand" to hold wood still. Otherwise, bouncing wood tends to defy the chain's onslaughts by falling off the buck, flying into your face, or jambing the chain to a stop. If your chain jambs in "small" wood, safe practice mandates that you shut off a gas saw or unplug an electric saw before attempting to *gently* jockey the chain free. If that fails, use a pocketknife to carve the chain out, rather than horsing it and risking chain damage.

Lacking a bucking vise, many backyard buckers depend on a helper to hold limbs steady. But this often brings the helper's hands and head dangerously close to the hungry chain. Lacking a helper, many sawyers attempt acrobatic acts, balancing on one foot while raising the other to clamp the wood in place. You will search in vain for

With saw idling in your left hand, use your right hand to place limbs into the buck. For your convenience, your supply of limbs and short logs should be near your right side.

With your left toe holding the limb vise lever down, and with your left heel firmly on the ground for good footing, you are ready to cut a stove-length log from the near end of the buck—the vise lever clamping the log in place.

a chainsaw manual that encourages one-footed sawing. (Of course, a resultant accident could leave one-footed sawing as the only option.)

When to Use the Vise. The two prime factors are wood weight and chain sharpness. If your chain is sharp, you'll need the vise only for long, skinny limbs and short logs of under 5-inch diameter that you want to trim to stove length. A dull chain can cause even the heavier wood to travel on you.

For heavier wood, you can remove the vise entirely and use the buck in the more traditional mode because heavier wood will stay put.

Materials. Buck parts can be fashioned from scrap 2 × 4s and 1 × 6s, three short dowels cut from an old broom handle, and forty-two 2½-inch No. 10 galvanized flathead screws. If you expect to leave the buck out in wet weather, either treat raw lumber with preservative or invest in pressure-treated lumber. Although screws result in a tighter structure over the years you might want to economize by substituting a few dozen galvanized nails. A nail-fastened prototype for the buck shown survived many tough cutting seasons. Over the years, the nails loosened a bit, but a periodic hammer blow here and there put the buck in shape.

Caution: Avoid substituting carriage bolts for the wooden dowels as pins for the crossing 2 × 4s. True, bolts are quicker to install, though more expensive. But such bolts would lie in the path of a misguided saw chain. A bolt could damage chain cutters badly. Worse, it could break a chain, with chain ends whiplashing at you. Or it could cause a vicious saw kickback. *Avoid bolts in the 2 × 4s!*

Vise parts consist of scrap 1 × 3; a three-inch long, 3/8-inch carriage bolt; two washers with a nut; two 3-foot lengths of nylon cord; and a few logs from your woodpile. Here, it's okay to use a bolt because your saw chain won't come within stovewood length of it if you use safe sawing practices.

HOW TO SET UP THE VISE ASSEMBLY

Prepare the vise's foot-lever assembly with a notch in the short fulcrum log and a notch in the far end of the long foot-lever log, as shown.

After laying the long foot lever into the notch in the fulcrum log, place the heavy ballast log into the notch of the long foot lever and then, as shown, lash it in place with cord.

To open the buck, fix one set of legs into the ground, as shown, while swinging the other set outward. Then lower the buck over the elevated foot lever so that the set of legs nearest you nearly touches the lever. With the buck in place over the foot lever, you can easily adjust the height of the foot lever by moving the fulcrum log in relation to the ballast log, or by moving the buck itself.

Use strong nylon cord to tie the vise lever to the foot lever. The several notches along the top of the vise lever allow quick minor adjustment of the cord for best clamping over the sawlog. You can make larger adjustments by changing the length of the cord or by moving the short fulcrum log, at ground level, closer to or farther from the biggest ballast log. Again, adjustments should allow your left foot to achieve solid footing—toe on foot lever, heel on ground.

Sequence continued next page.

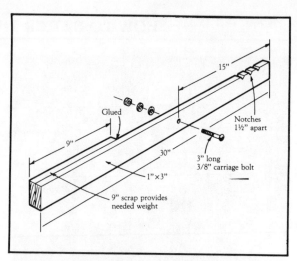

This shows the 1 × 3 vise lever in the at-rest position. Note that a scrap piece of 1 × 3 has been glued to the lever to provide enough weight to automatically raise the vise whenever cord tension is released. If the heavier end of the vise were longer to achieve the necessary counterweight, it could interfere with a helper's movements in a two-man bucking operation.

To mount the vise-lever parts, insert a 3-inch long, 3/8-inch carriage bolt through both the 1 × 3 and the 2 × 4. Washers promote a desired free swinging action.

The entire assembly is at rest, ready for bucking.

HOW TO BUILD THE BUCK

To mark the 2 × 4s for dowel holes, place four 3 footers and two 3½ footers side by side and score a line across all members 28 inches from the far end shown. Uniform height of the dowel holes is essential for folding action of the buck.

Center and then bore holes in mating pairs of 2 × 4s simultaneously.

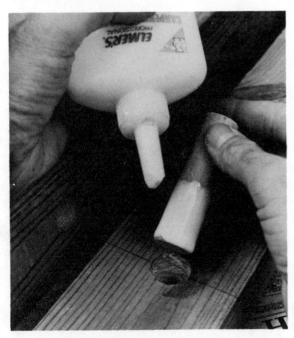

Glue half of each of the three dowels into one member of each mating pair of 2 × 4s. Use waterproof glue.

Make a final check to see that mating pairs mate. Then allow glue to dry.

Sequence continued next page.

After mating the paired 2 × 4s, space each pair by means of half a dozen playing cards at each end to ensure that the buck will open and close after fastening of horizontal 1 × 6s. For best functioning of the vise in conjunction with sawing, space the pairs of 2 × 4s so that each group of four, measuring from left and from right, measures about two inches less at the outside than the normal stove-wood length you'll be cutting. Before drilling screw holes, tack-nail 1 × 6s in place, checking carefully for spacing and square. You can then bore and fasten screws as you pull out nails. Important: Fasten 1 × 6s to the second, third and sixth 2 × 4s, reading from left. The drawing shows the lower two 1 × 6s staggered to make handsawing easier for a right-handed sawyer who wishes to step close over the work, at the right-hand end of the buck, opposite the chainsawing end. Accompanying photos show lower 1 × 6s staggered for a left-handed hand sawyer. If you don't plan to use the buck for handsawing, you can simply run one, rather than two, 1 × 6 across the bottom.

Before fastening the top 1 × 6, align its top edge exactly 1½ inches below the center of the dowel holes. That way the 1 × 6 prevents the buck from opening too far, should it be jostled.

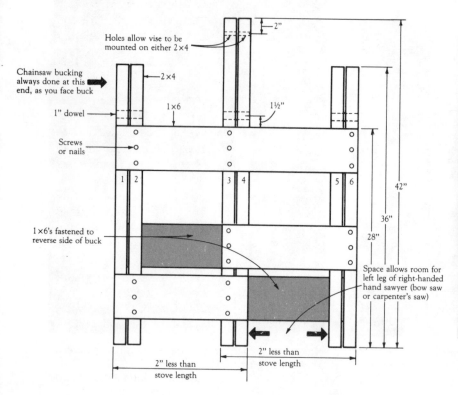

Holes allow vise to be mounted on either 2 × 4

2"

Chainsaw bucking always done at this end, as you face buck

2 × 4

1 × 6

1½"

1" dowel

Screws or nails

1 2   3 4   5 6

42"

1 × 6's fastened to reverse side of buck

36"

28"

Space allows room for left leg of right-handed hand sawyer (bow saw or carpenter's saw)

2" less than stove length

2" less than stove length

2" less than stove length

8

Bucking Piled Wood to Stove Length

Team bucking allows efficient, continuous cutting because the helper can place logs in the buck and then reposition them for successive cuts. Heavy logs will stay in place by virtue of their weight. But the helper must hold light limbs in place to prevent limb chatter and chain hangups on otherwise unstable wood.

For this "backyard" bucking, you'll need a sawbuck to support the logs at a comfortable sawing height. For efficiency's sake, the buck should allow you to saw at least a few stove lengths before moving the log or setting the saw down.

Team Bucking. For ultimate bucking speed on a goodly pile of logs, team bucking is the answer. Employing one saw and a partner or two to place logs on a couple of sawbucks, you can cut continuously, pausing only to reposition the saw and squeeze the throttle. In fact, one sawyer and a helper can buck through a cord of wood four to eight times faster than a lone sawyer.

Caution: Team bucking poses unique hazards both to helpers and the sawyer. A helper may be tempted to hurry too much in clearing away stove lengths and in lugging logs to the buck. In hurrying to keep up with the sawyer, he may stumble and fall or let a log smash his fingers as he lowers it onto the buck. Haste and zeal can also result in lifting injuries. In addition, a helper may reach in dangerously close to the saw bar. Here coincidental approaches of the saw bar and the helper's hands—Enough said!

To avoid cutting injuries, be sure sawyer and helper establish safety rules beforehand. The simplest and surest one forbids positioning of the saw for cutting until after the helper has placed and adjusted the log and given the sawyer a nod. Then the helper must never reach in to adjust the log until the sawyer has stepped back, removed the saw, and nodded a request that the log be repositioned. (*Note:* Communications can be mainly by nods and eye contact because your voices may go unheard with the saw running and while each of you wears ear plugs or muffs. Some plugs and muffs allow conversational tones to enter, while holding out high-decibel noise. Of course, even if you wear air-tight plugs, you can shout loud enough to be heard, but this tends to slow the work and signal the neighbors that your team needs more practice.)

The sawyer may occasionally signal the helper to turn the log or shift it lengthwise. Such moves can give the sawyer a better vantage over a crotch section or simply allow more efficient sectioning of the log.

Sawyer hazards? Of course, the sawyer must guard against kickback and other sudden hazards. There is a subtler one, however. The efficiency of team bucking tends to promote a rhythmic work momentum that urges the sawyer to cut continuously through a whole tank of fuel. If the saw has no vibration dampeners, or marginally effective ones, your hands and forearms absorb more vibration than is good for them. This may lead to an almost pleasant tingling sensation in your hands when you stop to refill the fuel tank. This tingling may fade even before you're ready to resume cutting. After another tank or two, though, your hands may feel hot and numbed enough to have lost their sense of touch. They may feel almost as though flesh, bone, and muscle have been transformed into rubber gloves filled with hot water. It's an attention-getting sensation, though usually not painful. Soon normal feeling returns. Hours later your hands may produce that strangely pleasant sensation again that is not quite tingling but almost. No pain. No worry yet.

It seems that sawyers vary in their susceptibility to these sensations that result from vibration trauma to nerves and small arteries. The arteries respond by constricting, thereby cutting off blood flow. If vibration becomes a daily thing, as is the case with professional loggers, arteries may become so highly sensitized that their constriction causes excruciating pain as well. The phenomenon is commonly known as white fingers. Doctors may term it Raynaud's syndrome or traumatic vasospastic disease (TVD). Many loggers simply endure the pain, which may come and go during sawing or even hours or days later. For loggers the problem can lead to an occupational disability. Surgeons have counteracted the disease by performing surgical bypass operations on the wrist arteries of those afflicted. *Doc's orders:* No more chainsawing!

Even the best saws, with the best vibration dampening systems, transfer some vibration to your hands. Your vulnerability to vibration injury depends on the amount of cutting you do, your sawing techniques, and your physical susceptibility. Best ways to avoid injury? (1) Use only saws with reasonably effective vibration dampeners. (2) Cut with the saw housing or bumpers on the log itself when possible so the log absorbs some of the vibrations. (3) Keep continuous cutting sessions brief by taking frequent breaks or by frequently exchanging sawing and helping duties with a partner.

One-man Bucking. If you're bucking up fewer than a dozen logs, you might as well mozy through the work. What's the hurry? Here you can place each log on the buck, buck off end sections, set the saw down, reposition the remaining overlength log section, pick up the saw, and make the final cut. Then you can shut off your saw, clear away the cut sections, and positon another big log. This process is safe and easy, though comparatively slow. It consumes a bit of your time, and it requires that you restart a gas saw frequently unless you prefer to waste gas instead, with the saw at idle while you police-up the stovewood and then go after the next log. An electric saw is more energy-efficient in this case. It runs only when you press the trigger.

For one-man bucking of small-diameter logs and limbs, you'll need a means of clamping the wood in place. Otherwise the saw chain tends to bounce off the wood, or cause it to rattle and twist. Here the chain also tends to jamb in the wood without cutting deeply and thereby bogs down the engine. For such bucking, consider using a cross-legged sawbuck *with a limb vise.* Its construction and use are described in the preceding chapter. If you instead want a helper to hold limbs in place on the buck, review preceding pages on hazards of team bucking.

The multiple-log buck in accompanying illustrations gives you ultimate one-man efficiency in cutting up a goodly pile of stovewood. It works well, provided the logs are heavy enough to stay put when the chain hits them. Depending on the density of wood species, this usually requires that short logs be at least six inches in diameter. The multiple-log buck lets you cut up a dozen or more logs into stove lengths in one series of passes before you have to set the saw down. Besides the efficiency of this buck, it costs nothing. You can find all components for it in your woodpile.

Once you have sawn your stovewood, you can save the buck's notched base logs for later use or cut them to stove lengths too. Then, during the following season, you can quickly fashion new components from your new pile of logs.

For one-man bucking of piled logs, this buck supports up to a dozen logs. It allows you to cut all logs into stove length in a single pass with no need to set the saw down after cutting each log as you would if using just the cross-legged buck shown in the background.

Three large-diameter logs serve as the "legs" here. Cut two notches on top of each large log so that parallel poles can be lain into the notches. Next, cut notches in the parallel poles so that logs to be sawn into stove length will be bedded securely. Space the notches about 1½ guide bar lengths apart to reduce the likelihood that the bar nose would accidentally hit the log in front, with resultant kickback.

Sequence continued next page.

After one relatively swift pass, saw logs like those at right have been reduced to stove length. Next set the saw down, toss the stove lengths into the splitting pile shown at left. Then place the next series of logs for bucking.

Logs of this weight will stay put in the notched pole until you make the final cut, as shown. At this point you may need to brace the saw log with your right knee. Note: In this case, be sure that a parallel pole separates your leg and the saw bar. Also to avoid back strain, avoid reaching far to your right. Instead walk around to the other side of the buck so that you can saw in a comfortable, balanced position.

9
Tree Felling Sanity

ENVIRONMENTAL IMPACTS

The energy crisis brought dramatic changes to the home landscape. Suddenly woodpiles and extra chimneys seemed to be everywhere. One good effect was a reduced consumption of fossil fuels in woodheated homes. Another effect was an enormous increase in chainsaw sales and wood-fuel consumption.

Aside from the bothersome noise a chainsaw creates and air-polluting effects of saws and wood-burning chimneys, the woodburning vogue can and has resulted in spottily abhorent impacts on the natural landscape. Thus, every chainsaw owner should be aware of the impact his sawing can have.

In suburban and semirural areas, you can harvest wood for minimum impact by taking only those trees that would be removed anyway. Here land clearing for roadways and real-estate development can give you a source of free or low-cost wood, or even wood that you may be paid to remove. The environmental culprit here is not the

The tenants in this snag tree give little thought to the insect-eating birds that helped excavate this two-door apartment nor to the cavity-nesting birds that once called this snag their home. People with chainsaws often fail to give this succession much thought too. (Photo by Irene Vandermolen.)

chainsaw, but population sprawl and a phenomenon that chambers of commerce term "progress."

Other nonrural wood sources include trees needing pruning and trees that neighbors want removed to make way for backyard projects. Such culling has minimal impact indeed. Another possibility is roadside deadfall.

Trees you might cut in rural or forest settings are less likely to be those destined for imminent demise. Here you have a wide range of choices—good and bad. Although wood is a renewable resource, remember that a 75-year-old tree will take about 75 years to replace. Yet properly managed, an acre of forest land is capable of yielding a cord of wood each year forever.

If you have a permit to cut on government land, you may be expected to take only cull trees, those that foresters have marked for removal either because they show poor potential as lumber or because they will shade out more-desirable sprouts. If you have free rein in a forested area, here are some thoughts on environmental impacts:

Save Some Snags. These are dead or dying trees. Many woodcutters mistakenly cut down all snags first, believing this cutting cleans up the forest. Snags are popular among woodburners because they often have seasoned dry on the stump and can be burned with little or no prior seasoning in the backyard woodpile. But some snags, at least, deserve protection.

The U.S. Forest Service prefers to have three to six snags per forested acre because snags play a crucial role in nature's chain of life. In well-managed forests, loggers are required to leave at least two snags per acre standing. Why? First, dying and dead trees incubate and nourish many species of insects, which in turn attract many species of insect-eating birds and even bears. Trees with cavities caused by rot and insects attract denning mammals as well as over 80 species of cavity-nesting birds such as owls, ducks, woodpeckers, swallows, and bluebirds. These birds require cavities for nesting, and most of them depend largely on an insect diet.

The insect-eating cavity nesters help restrict populations that feed on healthy trees. In fact, these birds are valuable in preventing insect plagues—so valuable that forest managers often attempt to attract the birds to areas with too few snags by mounting birdhouses to healthy trees.

Snag safety note: Snags can be dangerous to cut because they may have dead branches, "widow makers," that can fall on you if shaken loose by vibrations or impacts made against felling wedges. These trees are also known to break in sections and send branches flying in many directions as they begin to fall or hit other trees.

Nut Trees. Many of the most highly prized woods for cabinetmaking, tool handles, and firewood come from nut trees. Here the strength of these woods, their beautiful grains, and their high density conspire to bring them down. Yet nut trees—especially the oaks—are among the most valuable trees to wildlife. Whole population cycles of squirrels follow the annual size of the mast (nut) crop. When acorns are dropping, bears will even abandon the bounty of garbage dumps for the pleasure and nutrition of acorns. Deer are highly dependent on the nutrition and fat reserves a good mast crop can provide for winter. Waterfowl like acorns too.

Many other trees, such as fruit and berry trees, also play important ecological roles. In short, many of the trees that may look best to you for your own purposes—whether for their proximity to roadside, straight easy-splitting grain, or high Btu value—may be better left standing. Before you select trees for cutting, try to plan for the least impact, even if this requires some sacrifices of you.

Winter Browse and Wildlife Cover. If you are planning to fell numerous trees in an area, consider felling them in early winter. The branches you leave behind will provide nutritious browse for deer and rabbits. Northwoods loggers claim that the sound of their chainsaws often serves like a dinner gong for hungry deer. Come spring you can heap up the uneaten branches to serve as home and cover for small wildlife.

STARTING SMALL

Before attempting to fell large trees, be sure to give yourself plenty of practice on small and medium-size trees, working your way up in size as your experience *and equipment* allow. Be especially reluctant to tackle any tree with base diameter greater than the length of your saw bar. Such trees require multiple cuts of high precision. If such large trees also contain trunk rot or if they lean heavily to one side, stresses within the trunk may lead to dangerously bizarre falls long before you anticipate. More on this a bit later.

Small-diameter hangers like this red maple regularly injure woods workers and damage saw bars. Embarrassed and perhaps angry, many workers mistakenly go after hangers with their chainsaws.

SIZING UP THE TREE

For most of us, the six most critical factors are (1) the tree's size, (2) its lean, (3) the condition of its trunk and branches, (4) trees in the path of the intended fall, (5) the wind factor, and (6) the escape routes. If you are felling trees for sawmilling, you must also consider boulders and terrain features that could result in trunk breakage as the tree hits the ground.

Direction of Fall. Often, the easiest and safest option is to fell the tree the way it leans. Lean is determined by the amount its trunk leans from vertical as well as asymmetry of the limbs. On the other hand, nearby trees might catch a tree falling toward its natural lean and thus leave you with a *hanger.* If you decide not to take the tree the direction it wants to go, you may have to employ a number of compensating measures. First, you could "top" the tree by climbing up and using handsaws to remove limbs or whole tree sections that weight the tree away from the direction you want it to fall. Second, you might employ heavy-duty pull ropes secured to mechanical pullers. Third, you might force wedges or a pry bar into the backcut, behind the saw; these can appreciably lift the tree away from its natural lean. Fourth, you might make an unequal hinge cut, shown in later drawings, such that the wider side of the hinge wood brings the falling tree slightly toward the wider side of the hinge. And you might employ combinations of the above.

Hazards. Tree size includes the spread of branches as well as height. If a falling tree's limbs strike those of other trees on their way down, bad things can happen: (1) The tree can hang up in the limbs of other trees, leaving you with a dreaded hanger. A hanger may jump its own stump or hold precariously to the stump, exerting tremendous stresses there. A hanger can also exert enormous stresses on the supporting tree and its limbs. (2) The falling tree can also twist severely, its base jumping wildly to either side or backward from the stump. (3) Then too, colliding limbs can snap,

Here Daniel Tilton, instructor in Scandinavian logging methods, demonstrates how to twist the trunk of a small hanger by means of a pry bar with cant hook so that upper branches rotate off the branches of the supporting tree. Caution: This technique should not be attempted on larger trees. And you must first cut away hingewood on the side of the hung tree nearer the supporting tree and push the pry bar from that side too. Otherwise the falling hanger could pin you under the bar. Of course there's danger that the hanger could jump toward your feet; but the lower the stump, the lesser the chance. Who said small trees are easy?

This white pine was intentionally hung up to illustrate use of a 10-foot pry pole, here consisting of a sturdy birch sapling. The pine was too firmly hung to be twisted free. So Tilton first cut away remaining hingewood and now prepares to lift-skid the trunk toward the camera.

Several lifts with the pry pole moved the pine toward the camera about 15 feet until hung-up limbs broke free. You can see the original stump between the downed white pine and Tilton's thigh. Caution: This method is strenuous, and there's danger that the falling tree could whip the pry pole over you unless you step aside in time. Pull ropes partway up the hanger may be safer in many instances.

rocketing in any direction, perhaps striking you even though you may have retreated a relatively safe distance from the stump.

Bad things 2 and 3, above, should cause embarrassment if they don't cause injury. But bad thing Number 1, the hanger, leaves you both embarrassed and perhaps primed for total disaster. The most dangerous options are to attempt cutting near the hanger's base or to climb the supporting tree to make freeing cuts, or (*Watch out!*) to attempt to fell the supporting tree. Woodcutters are injured or killed regularly when attempting such hazardous measures.

In the woods, the safest measure is usually to leave a firmly hung hanger to nature's course. Yet if a hanging tree would later present a hazard to people, or if you must have it for its wood or your ego, consider means of twisting, lifting, or pulling it off the supporting tree. Logging companies often attach cable to hangers and pull them free with heavy vehicles or mechanical pullers. You may find that a heavy-duty rope or two and mechanical pullers will do the trick. Or you might employ a wooden pry pole or a steel bar with cant hook. *Final note:* Since a hanger almost always results from an incorrect assessment of the situation or from imprecise felling cuts, it should make you especially mindful of avoiding similar mistakes in the future.

Condition of Trunk. Here are some checks: Dead or dying branches in the tree's upper reaches sometimes indicate trunk rot. Cavities in the upper trunk may be tipoffs too. A surer check is to whack the trunk with the heel of an ax. If you feel and hear solid wood all around the trunk, the trunk will probably be sound. But if you hear hollow reverberations, you can count on extensive heart rot. If your ax hits punky, mushy wood, this tree could be trouble. All rotted trunks make for dangerous felling because of the uncertain amounts and strengths of available hingewood for the felling cuts. Before tackling any tree with trunk rot, remember that dead and dying trees play important roles in nature's life cycle. If such tree is large and poses no immediate threat to people or property, consider leaving it alone.

Sequence continued next page.

FELLING-CUT DIAGRAMS

The accompanying drawings illustrate the simpler cuts, as well as complex cuts that should be attempted only by highly skilled fallers. Skilled fallers will tackle trunks whose diameters are more than double the length of the saw bar. But this calls for the eye and feel that come only through experience. If you want to fell a tree whose trunk diameter approaches or exceeds double the length of your saw bar, first practice needed cuts on tall stumps until you've mastered them. Or consider these options: Hire a pro with a big saw just to fell the tree; then you can buck and limb it. Or rent a big saw for the project. Or you might buy or borrow a hand-powered two-man crosscut saw.

A handsaw calls for hard labor from two men, yet it gives you full-diameter lines of cut at precise angles, and handsawing lets you see the widening kerf of the backcut and hear the first cracking sounds in the trunk when the treetop begins inching in the direction of the fall. Handsawing also lets you better hear warning shouts of treetop movement made by a safely distant observer. So with the first cracking or the first shout, you and your partner can pull the saw, drop it there, and make tracks.

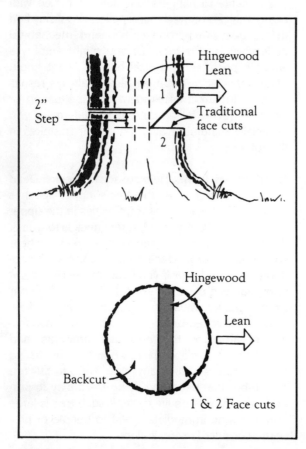

BASIC FELLING CUTS

Traditional American cuts include face (pie) cuts that meet at about one-third trunk diameter and form roughly a 45-degree angle. The backcut is horizontal and about two inches higher than the intersection of the face cuts. The backcut should penetrate only far enough to allow an inch of remaining hingewood on small trees and two inches on larger trees. (Some fallers mistakenly keep cutting the backcut, right through desired hingewood, until the tree finally falls —very dangerous. The elevated backcut provides a step (stump shot) that helps prevent the butt of the falling tree from jumping off the stump as upper branches impact against branches of standing trees. The hingewood serves, hinge-like, to control the tree during the fall and helps to prevent its butt from twisting off or jumping the stump. Note also that the upper face cut is made first. Its kerf then lets you see your chain the moment it intersects during the lower face cut, and this tells you to stop before mistakenly overlapping into hingewood.

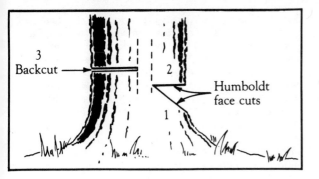

The American Humboldt cut is often used in professional logging on large-diameter trees for these reasons: (1) To conserve the maximum amount of wood above the cuts, (2) to allow the butt of the tree to slip to the ground before the upper section strikes the ground and thus minimize chances of trunk breakage, (3) to help prevent the butt's jumping backward off the stump, especially when felling a tree uphill. As for the traditional American cut, the backcut here is horizontal and about two inches higher than the intersection of face cuts—to provide the safety step.

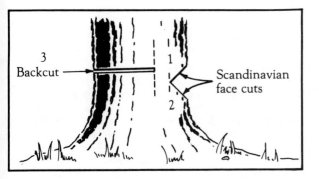

Scandinavian face cuts are made at about 90-degree angles so that the faces don't close until the tree hits the ground. Thus, the butt of the falling tree remains attached to the stump throughout the fall—hingewood guiding all the way. This greatly reduces the chance of fiber pull. That is, with the American traditional 45-degree face cuts, long wood fibers are often pulled from the butt of the falling tree because the faces close too soon and thereby pull the fibers; this lowers the market value of the butt log. Note here that the elevated, horizontal backcut ensures a step during the fall that helps prevent the butt from jumping backward off the stump should the tree's upper branches impact against branches of standing trees. The 90-degree face cuts also help prevent stump jump.

There are further distinctions between Scandinavian and American felling cuts. Although the 90-degree Scandinavian pie reduces the likelihood of fiber pull and stump jump, it also increases the chance that wet hingewood, especially in spring, will bend over all the way without breaking off. You must then cut off the hingewood, which may now be under severe stresses capable of making the freed butt jump back at you or roll violently to either side. Note also that Scandinavian face cuts tend to be a little shallower than the American cuts. Here the theory is that the better control fibers for the hinge are in the sapwood, rather than in the heartwood. Also, on trees leaning in the direction of the face cuts as well as on perfectly vertical trees, the farther the hinge pivot point is from the backcut side of the tree, the greater your leverage when driving wedges or using a pry bar to lift the tree in the direction of fall.

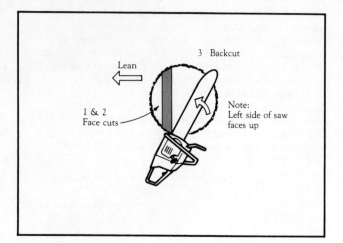

Simplest, safest felling cuts are made on trees with stump diameters no greater than saw-bar length. The goal is a parallel strip of hingewood up to two inches wide on large trees. Note that the left side of the saw faces upward for all three cuts. This allows best control and keeps you aside from the saw's kickback zone. Should you fail to make the face cuts meet precisely, trim them to perfection before beginning the backcut. Cuts 1 and 2 must not overlap at the sides, and their intersection should be at a right angle to the intended "lay" of the tree. It's often wise to borrow or rent a saw with a bar as long as the trunk diameter if you've not mastered the other option: making multiple cuts with a bar shorter than trunk diameter.

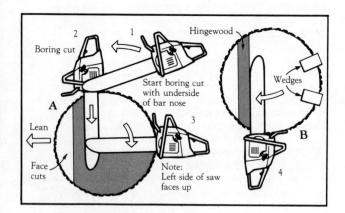

Trunk diameters more than bar length but less than two bar lengths: Don't attempt these felling cuts on trees until after you've mastered them in practice sessions on stumps. Use a crayon to draw guidelines on the bark. Again, make the upper face cut first so that you can see into the kerf when completing the lower face cut. Then trim up the face cuts to your satisfaction before making the backcut, which begins as a boring cut about two inches higher than the intersection of the face cuts. After you've progressed halfway around the backcut, jamb a plastic wedge into place by hand. Cut a little farther and then pound the wedge deeper to prevent the kerf from tightening on the bar. Continue the cut until you can insert a second wedge, driving both wedges in only so far as to allow safe clearance for the chain. Further wedge penetration can lift the tree appreciably toward the intended fall. Always use two wedges, especially in winter, in the event one squirts out like a pinched pumpkin seed.

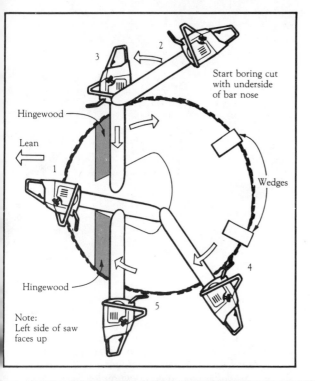

3
2
Start boring cut
with underside
of bar nose

Hingewood

Lean

1

Wedges

Hingewood

5

4

Note:
Left side of saw
faces up

This shows the cuts pros use to fell trees up to 2½ saw bars in diameter. Here it's important that the boring cut be made in the upper face and into the heartwood as near as possible to the height of the backcut to come. The boring cut should allow plenty of hingewood on either side. Note: For safety and control, pros keep the left side of the saw nearer to them than the topside.

RESIDENTIAL FELLING

Many woodcutters prefer this to woods work, especially if the trees are close to home, allowing a short haul of the wood. If you have a saw, neighbors may ask you to do the cutting, perhaps working alongside you and sharing the wood. Then again you may be offered payment for your services.

If you're just helping a neighbor without pay, your homeowner's insurance policy will likely cover you for damage you might do. But if you agree to work for a fee, you technically become a professional, liable for damages, and you may also be in violation of local ordinances. Some localities require that pros be licensed and specially insured for tree work. Insurance organizations recommend that you cover yourself with comprehensive insurance of at least $100,000.

That said, there are other good reasons to decline residential felling jobs—reasons you don't have to worry about in the woods.

Spectators. People seem compelled to watch tree work, often at unsafe distances. Yet even if you manage to make everyone stand back a safe distance and even if you drop the tree exactly where you want to, tragedy can result. For example, a pet excited by initial cracking of a falling tree might scamper wildly into the tree's path, with its master or mistress chasing frantically to retrieve it.

Property. Many a house, garage, fence, and vehicle has been demolished by an errant-falling tree. Besides, even a small limb falling from 60 feet can render great damage. And there are hidden property items too, made known after a fallen tree has broken water pipes or septic lines that the owner once swore lay somewhere else.

Nails. Residential trees often hide large nails overgrown by bark and tree rings. Such nails will lie exactly at the height at which they were driven. Many people mistakenly believe that growing trees lift buried nails each year. A big nail can irreparably damage saw chain cutters and even break the chain, causing it to whip wildly.

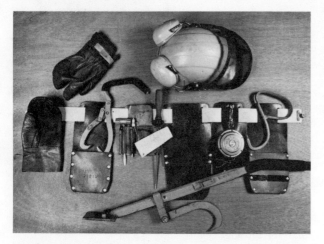

Scandinavian gear gaining acceptance in North America includes mitts with ballistic nylon backing and a right-hand trigger finger plus a gear belt weighing 6½ pounds when loaded. The belt shown holds a lifting tong, a scrench, a carburetor screwdriver, a plastic wedge, a round file, a measuring tape, and a piling hook. The spring-steel bar at bottom can serve as a felling lever, log pry, log twister, and log carrier. The hardhat has ear muffs and a face screen.

Essential tools for traditional American felling include the saw, a hardhat, some type of ear and eye protection, large and small wedges, a hatchet for driving wedges and trimming gritty bark, and a wire brush. Use the brush especially on smooth-barked trees on which rain splash has deposited fine grit. If you doubt the brush's value, watch competition lumberjacks brush down logs before chainsawing, handsawing, and chopping events. Even airborne particles deposited on fallen logs in a rain will dull your chain.

PULL ROPES AND SAFETY ROPES

Various half-inch manila and synthetic ropes are rated with breaking strength over 2,000 pounds. To ensure adequate margins for safety, though, never tax a rope beyond one-seventh its breaking strength. Thus, half-inch manila should never be stressed more than a few hundred pounds.

Yet even a few hundred pounds pull can exert tremendous leverage if you secure the rope high up in a tree. Used in combination with wedges in the backcut, a pull rope can be a useful tool indeed.

Caution: Avoid using a rope and vehicle to pull a tree away from its natural lean. Even though the vehicle may have power enough, the rope may not have strength enough. Many is the pull rope that snapped when the tree decided it would go nature's way rather than the vehicle's way.

Still, judicious use of a pull rope in league with heavy safety ropes and wedges in the backcut can, to some degree, correct for an undesireable lean. Here it's important to "take up" and re-secure the safety ropes each time you "lift" the tree with wedges and then exert a few hundred pounds pull with the pull rope. As to length, the ropes should be longer overland than the reach of the tree. This will help ensure the safety of everyone taking up on the ropes.

SCANDINAVIAN FELLING OF TREE WITH TRUNK DIAMETER GREATER THAN TWO SAW BAR LENGTHS

Daniel Tilton surveys a large, dead white pine before felling it with Scandinavian techniques. Whether you employ Scandinavian or American cuts, close study of the following photo sequence will give you pointers of value for either approach.

Before cutting, Tilton determines where he wants the tree to fall—based primarily on lean and clear space for the fall. He sights along the saw's falling guide marks (perpendicular to the saw bar) directly to the intended lay and adjusts saw position accordingly. This step calls for much the same kind of eye and feel as is needed for golfing.

Sequence continued next page.

Leaning into the tree and with the saw's back handle braced above his knee, Tilton begins the upper face cut.

Tilton continues the upper face cut from the opposite side of the tree, kneeling as necessary for control while keeping himself aside from the bar's kickback zone.

Tilton makes the lower face cut by drawing the bar along from right to left, inching backward with his feet. Note that his right arm is braced against his right thigh. One secret for making a precise intersection of the upper and lower face cuts is Tilton's sighting through the kerf of the upper face cut. The other secret in making this difficult cut lies in the disclike bulge in Tilton's right thigh pocket. It's caused by a can of Skoal chewing tobacco.

Having removed a 90-degree pie, Tilton pauses to compare the face with the intended direction of fall. This is the time to spruce-up the face cuts if they're not precise. If you've held the tobacco in your lower lip, it's also the time to spit—first lifting the face screen.

Sequence continued next page.

When the trunk diameter is more than twice the length of the bar, experts bore into the heartwood at the same height they will make the backcut—here about 1½ inches above the intersection of the two face cuts. This bore cut must not be so wide that it cuts needed hingewood.

Tilton begins the backcut 1½ inches above the intersection of the face cuts, using the bottom of his bar, before boring straight in. Always begin boring cuts with the bottom front of the bar in order to avoid unruly handling or a kickback.

After walking the saw more than halfway around the backcut, let the saw idle there and insert the first wedge.

If Tilton doubts that his backcut reached as far as the boring cut in the upper face, he cuts a side notch as shown and works the bar in deeper.

Sequence continued next page.

Using American methods, you would insert a wedge here. Instead, Tilton inserts a deflated nylon air bag with pressure hose that connects to a pressure nozzle on his Jonsereds saw. By regulating the flow of air, Tilton can synchronize inflation with his final cut. One advantage here is that you can continuously increase lift as your backcut approaches the hinge. Wedges require that you stop and whack periodically, while leaning stresses within the trunk could begin to split fibers vertically up the tree, perhaps causing the dreaded barberchair. On the other hand, if the chain exploded the bag, the tree would sit back suddenly—the inertia possibly tipping the tree over backwards.

After a successful fall, Tilton has here replaced the air-bag and the saw to simulate their working relationship. Note the evidence of the bore cut made earlier in the face.

THINNING

In thinning a stand of trees for best timber yield, remove the weaker or less desirable of two stems emerging from one trunk. After making the face cuts here, Tilton made the backcut by boring to form the hinge and then cutting toward the second tree. Here boring was necessary because there was no room to begin a conventional backcut. The boring backcut is also used when a tree leans heavily toward the face cuts. In this case the holding wood on the backside is last to be cut and helps prevent the tree from falling too soon. Otherwise a heavy leaner can cause a barberchair splitting of fibers up the core of the tree and a wildly misplaced fall.

Sequence continued next page.

With the less desirable of the two stems felled, cut a sloping top on the stump. This promotes rain runoff and lessens chances of damp rot and ice damage. It is likely that the remaining tree will grow large and healthy now, nourished by a root system that formerly fed two stems.

If you wish to fell both trees, form a notch on the second stem that will guide the second stem over the first, as shown. Of course it's wise to limb the first felled tree before felling the second tree over it.

Last, remove the stump, which in this case will make excellent though tough splitting firewood. Several cautions: *Rain splash can leave chain-dulling grit embedded in low-lying bark, especially on deciduous trees. Here use an ax to remove gritty bark, or cut at least eight inches above ground level. On this double stump, as on most stumps that formerly served two or more trees, the heartwood serving both stems was rotted and was filled with wind-blown grit that had worked its way between the trees and eventually into the heart of the stump. This hidden grit dulls chain fast. Here Tilton walked the saw bar around the stump in one continuous slice. In such a case you must drive wedges as you near completion to prevent the settling stump from pinching the saw bar.* Stump note: *Logging companies usually require that stumps be cut low enough to resemble manhole covers; this allows vehicles to drive over and it eliminates tall stumps that could break trunks of trees felled over the stumps.* Aesthetics: *Stumps cut near ground level are less conspicuous and decompose sooner, imparting a less disturbed look to the forest sooner. But if you vow you'll cut all stumps low, you'll be filing chain oftener.*

To make felling cuts on smaller trees, you get best saw control by laying your left shoulder against the tree as shown, spreading your feet, and bracing your right elbow against your right knee. Avoid cross-hand grips, off-balance crouching, and kneeling positions that put any part of you in the bar's kickback zone.

Here Tilton uses a pry bar to dump a small tree. His legs do the lifting. With such a bar the average man can deliver about 3,300 pounds of lifting force and apply sustained leverage as the tree begins to go. Note that Tilton here watches the tree's top section for movement and to spot any broken limbs that might fly his way.

Felling cuts on these poles indicate that the fallers wanted the billboard to go the other way. The sloping backcuts served no advantage. In fact, horizontal backcuts would have allowed use of wedges and increased the likelihood of fall toward the face cuts. The billboard shown was one of three felled one Saturday night near Nashville, apparently in retaliation for illegal cutting of trees that had begun blocking motorists' views of billboards. Similar feuds have erupted in other states. In New Mexico, a state spokesman once claimed that "roaming vigilantes have for the past 10 years effectively kept many roads almost completely free of billboards by using chainsaws, axes, and fires." Note: Billboard felling is illegal, and it is extremely dangerous work. Special hazards include electrical systems, the effect of wind on the board, and the probable need for simultaneous backcutting of several dry and brittle poles. Anti-billboard sentiments may better be served through legislation. (Nashville Banner photo by Vic Cooley.)

10
Limbing and Bucking Felled Trees

LANDING THE TRUNK OFF THE GROUND

The symmetry of branches on most conifer trees usually keeps the trunk of the felled tree well elevated for easy bucking. But long-trunked deciduous trees often make it advisable that you first analyze terrain, and if necessary, prepare the ground to receive the trunk. *Note:* If you are planning to save trunk sections for milling, you'll want to avoid hitting large boulders or cross-lain logs that could shatter the trunk when it hits the ground. If you are merely cutting firewood, trunk breakage is less important.

In either case, your goal should be to ensure that the felled trunk is supported by several

As to saw size and weight for limbing and bucking, there are two conflicting views among professional loggers. One view is that lightweight saws with short bars, as shown here, are less fatiguing to manhandle and help you keep an eye on the kick-back-prone nose of the guidebar. The other view is that saws with longer bars require less stooping and bending and thus impose less fatigue, and if a long bar is mounted with an antikick nose guard, the chances of rotational kickback are eliminated. In this photo, Daniel Tilton pauses with a lightweight saw after limbing and bucking the white pine he's shown felling in the previous chapter.

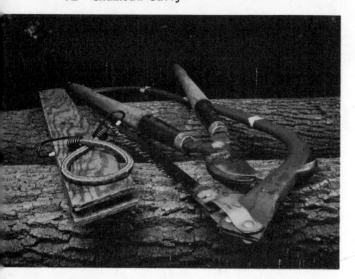

Lopping shears and bowsaws are far safer and almost as fast-cutting as chainsaws. They seldom stall and always start. They are quiet and emit no fumes. They let you take your time, never worried about wasting fuel at idle speed. In short, they have many advantages over chainsaws, even before you consider the chainsaw's kickback hazard unless the saw is equipped with a bar nose guard. Quality bowsaw blades cost no more than one professional saw chain sharpening, and the high-carbon blades made in Scandinavia seem to last forever. You can protect blade teeth and avoid injury from them by making wooden blade guards held in place with shock cord as shown at left. Consider using a 42-inch bowsaw for most work, even for light bucking. If you jog regularly, your heart and lungs will probably be up to serious bowsawing. Pushups and pullups can help ready the rest of you.

fulcrums. These keep the bark free of chain-dulling grit, reduce the chance that the chain will contact the ground, and give you several fulcrums at which you can overbuck straight through.

If you know that a long-trunked deciduous tree will destroy smaller trees and saplings as it falls, consider felling the smaller ones ahead. Otherwise they could become spring poles, bent under the large tree, and often hazardous to cut. Especially hazardous are the spring poles that you mistakenly believe to be limbs emerging from the bottomside of the felled tree. In this case, what you expect to be a routine limbing cut can turn to dynamite.

Lacking support legs, you might be able to take advantage of a few high spots along the ground. The least desirable lay is one in which a long heavy trunk is supported only at opposite ends. This results in a sagging sort of stress of the trunk that makes bucking difficult—even treacherous.

TOOLS FOR LIMBING AND BUCKING

Logging companies teach their buckers work-flow patterns that give *fastest limbing sequences* with reasonable regard for safety. This chapter will instead cover *safest limbing sequences and tools*, with reasonable regard for efficiency.

Lopping Shears. For limbs up to 1½-inches in diameter, heavy-duty lopping shears are about the fastest and safest tool. Such shears can handle nearly all limbs on medium-size conifers and most limbs on large conifers. Shears can also tackle most limbs on deciduous trees that would otherwise vibrate and bounce dangerously during chainsawing. A set of shears works well whether you are lopping to produce poles or to produce stove-length kindling or even mulch. In fact many wood-burners waste such limbs, leaving them behind because they are too dangerous to tackle with a chainsaw. With a shears you can make cuts just about as fast as you can open and close the handles—and that's fast.

Handsaws. Pruning saws and especially bowsaws can handle much work on which you might use a lopping shears—and then some. Slightly slower-cutting than shears on limbs to 1½-inches in diameter, handsaws make quick cuts on limbs up to four inches and more. Handsaws are also the safest tools with which to cut potentially dangerous spring poles. Here they let you grasp with your free hand the weaker portion of the pole that would otherwise spring upon being cut free. If you attempt to cut spring poles with a chainsaw, either part can spring free. Here ends might fly up and injure you directly or whip your chain bar as violently as if the saw chain had caused a kickback. Handsaws also come in handy to help buck free a chain-saw bar pinched in a saw kerf if a pry bar fails to do the trick. (Such pinches always follow carelessness or incorrect assessment of the stresses on a log. After freeing the saw bar, consider causes for the pinch in order to avoid the problem in the future.)

Ax. As a weapon, an ax can be a formidable ally. It has often been used to fend off charging bears. With hard work, an ax will also fell mighty trees. Yet if you own a chainsaw, an ax becomes a secondary cutting tool, though it can be highly useful. Here a single-bladed ax serves better than a double-bladed. It's good for these tasks:

• Sounding trunks for rot, using the ax heel
• Driving plastic and soft-metal felling wedges into a trunk's backcut
• Stripping gritty bark away so that saw teeth won't be dulled
• Lopping lower limbs from standing conifers to provide chainsaw-cutting clearances
• Limbing felled conifers (In most cases, you should work from the felled tree's base toward the top, trying to keep the trunk between you and the ax blade as much as possible.)
• Fashioning and driving wooden felling wedges and also broader wedges that can serve as jacks under felled trunks.

Lifters. Log lifters are useful for turning logs and lifting ends. Those designed only to elevate a log end for bucking are less versatile than magazine ads reveal. True, a lifter designed for bucking will elevate one end of a log. Remember, the principle is to lift and prepare for sawing less than half of the log, by weight. Once you have cut near the lifter, you must reposition it.

Do-it-yourself Lifters. You can elevate large-diameter logs by driving wooden wedges under them. For this you can use the wedge you removed from the face cut when you felled the tree, as well as an assortment of wedge-shaped blocks or limb sections. For logs you can lift at one end, it's often efficient to throw them onto another log and cut the elevated ends. In fact, if you are planning to buck a pile of logs, you can prepare more

Log lifters such as these have limited value. At most, they let you buck stove lengths from less than half a log before the need to reposition the lifter or wrestle the log by hand. You can usually do as well by propping the log to be bucked upon another log.

than half of them for bucking at one time by simply laying them over one another. Once you've made the first sweep with your chainsaw, you can then remove stove length sections and place the former support logs on top of remaining sections. After that sweep, remaining logs will be light enough for lifting onto saw bucks.

Chainsaw. Unless your saw bar has an antikick nose guard, it's usually safest to use your chainsaw only on tree sections you cannot handle efficiently with hand tools.

Accompanying drawings show recommended cutting sequences that take into account stresses on limbs and trunks for the situations shown. Each tree becomes a new problem for stress analysis, and these stresses surprise many buckers who haven't carefully considered effects of points of support under the tree.

In addition here are special cautions:
- Avoid using a chainsaw on unstable limbs or on spring poles.
- Use an antikick bar nose guard for limbing if possible.
- If your saw has no nose guard, avoid causing kickback by inadvertently touching the bar nose against an obstruction. Be especially wary of kickback that can result while the chain is coasting to a stop.
- If your saw has a chain brake, engage the brake on an idling saw before you walk with the saw or scramble over felled wood.
- Consider how trunk sections will roll once they are cut free. Often ground slope lets a free log roll downhill, *but not always.* Supporting limbs can exert a twisting stress on the trunk, causing it to "fall" uphill. Twisting stresses are often difficult to gauge.

Firewood cutters often mistakenly cut off whole limbs near the trunk and then chase around after them on the ground—letting the chain torment the dirt a bit. With hundreds of chain cutters passing any one point on the bar per second, it takes only a fraction of a second to thoroughly dull a chain. It's better to trim limbs back with loppers or bowsaws until they are stable enough not to chatter under a chainsaw. Then saw stove lengths right back to the trunk.

When using a chainsaw to limb conifers, try to keep the trunk between your legs and the chain. Whenever safe and practical, rest the saw housing on the trunk and saw with a rocking motion—transmitting all saw weight and vibrations to the trunk. If your saw bar has no antikick nose guard, keep an eye on the bar nose and be ready for kickback constantly. Avoid letting an unprotected bar nose blunder into limbs. Be especially wary of mistaking a spring pole for a tree limb.

Here Daniel Tilton places his fingers on four points at which he will attempt to weaken spring-pole stresses by making cautious shallow nicking cuts with his chainsaw. He'll begin cuts where his index finger is and work downward. In this situation, bowsaw cuts are safer for the nonprofessional. Even so, the spring pole could snap and lash upward at any moment, whipping the saw with it. Tilton likens spring-pole work to dismantling a boobie trap.

SOLVING BUCKING PROBLEMS

Chainsaw bars are often pinched and bound solidly in saw kerfs because the sawyers fail to anticipate internal log stresses. Then bars and chain are often damaged because sawyers attempt to free them by yanking on the saw handles; upcoming drawings show better ways. It's important to understand that logs supported at both ends are under a sagging sort of stress that compresses wood fibers on the topside and tensions them on the bottomside. If supported at only one end or directly under the intended cut, the wood fibers on the topside are tensioned and those on the bottomside are compressed. Analysis of stresses is further complicated when a log's supporting limbs exert twisting stresses. Here are some common problems and their solutions:

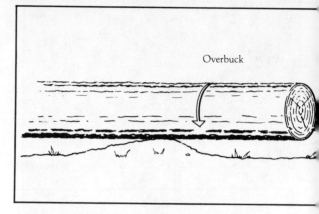

Use a simple overbuck to cut stovewood lengths when the log is supported on the side of the cut opposite the log you'll cut off.

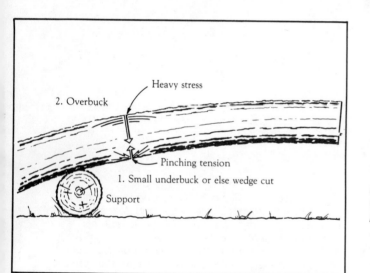

To remove a longer section under stress, first underbuck or make a wedge cut underneath and then overbuck. Otherwise the stress could split the unsupported log.

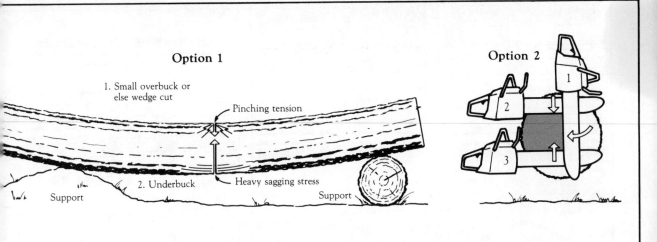

Option 1

1. Small overbuck or else wedge cut

Pinching tension

2. Underbuck

Heavy sagging stress

Support

Support

Option 2

If a log is under a sagging stress, you have two options. First, you can overbuck and perhaps cut a wedge topside before underbucking. Second, you can cut the backside before underbucking and then overbucking.

To section a long log supported at both ends, you might insert a support log and then overbuck partway down before underbucking at the angle shown. The angled underbuck makes the left log section fall away from the saw bar—the support log preventing the simultaneous collapse of the logs.

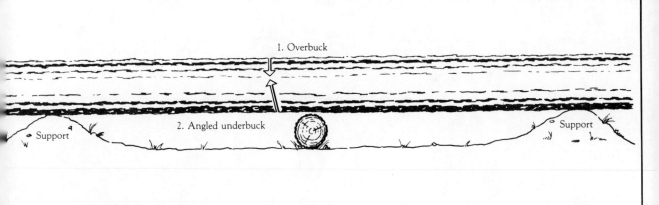

1. Overbuck

2. Angled underbuck

Support

Support

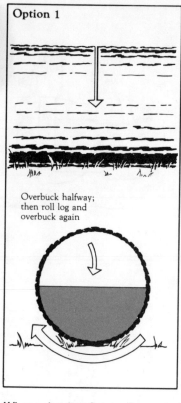

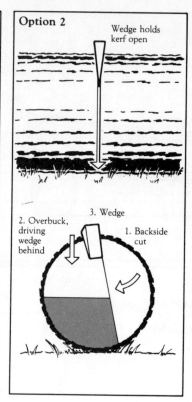

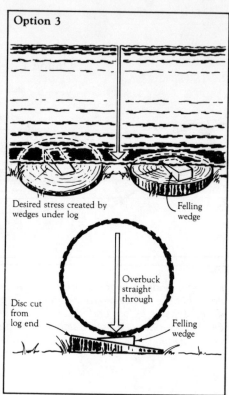

When a log lies flat on the ground you might use any of several options, to prevent chain from contacting dirt, depending on the log diameter and positions of remaining limbs. Option 1: *Overbuck partway, roll the log, and then overbuck the other side.* Option 2: *Cut the backside and then follow an overbuck with a wedge to keep the kerf open; remove the bar before cutting through the small hinge of wood near ground level and then twist the logs apart.* Option 3: *Elevate the log near the intended cut by driving a wedgeshaped disc (cut from a log end) underneath. If necessary, you can further elevate the log by driving a felling wedge or two between the disc and the log. With the wedges lifting, the stretched fibers topside and the tensioned fibers bottomside will keep the kerf open for an overbuck straight through.*

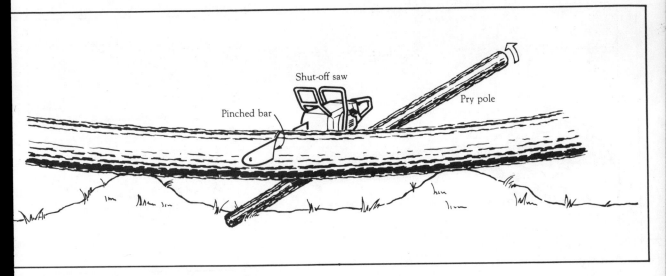

Many wood cutters mistakenly attempt to use a second chainsaw to cut a pinched saw bar free. This often results in a second pinched bar or chain damage when the second saw's chain hits that of the pinched bar. Usually it's better to find a way to open the kerf by changing stresses on the log. On small logs a pry pole, consisting of a sturdy sapling or limb, may do the trick.

To relieve kerf pinch from a large log, try to force a limb, log, or wedge-shaped disc underneath. Then drive felling wedges, if necessary, to open the kerf.

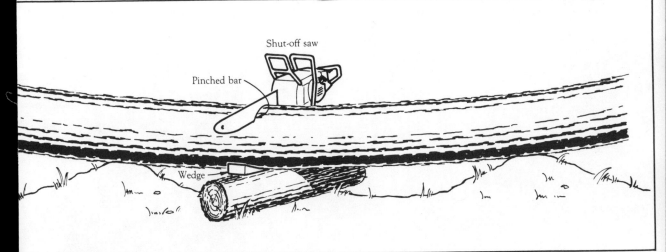

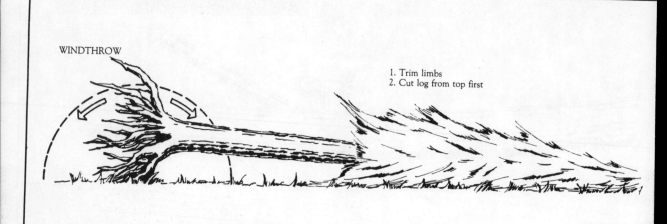

WINDTHROW

1. Trim limbs
2. Cut log from top first

Windthrown trees with roots still substantially anchored can be treacherous to buck because the roots may still be resilient enough to whip the trunk vertically once you cut away some of the tree's top section. Preceding drawings will help you determine bucking cuts on a windthrow's upper section. Carefully made cuts on the lower trunk may need be wedge shaped, depending on stresses. Be ready in case the stump suddenly whips upright. And never stand under upturned roots.

11
Lifting and Moving Logs

Brute Force vs. Good Sense. Usually, if you can hoist the entire weight of a log, the fastest way to transport it to a vehicle or woodpile is to hoist and carry it, even if you must stagger a bit. Yet if hoisting and carrying would be tough going, reconsider.

Such exertion may lead to pinched nerves in your joints or to pulled muscles. If you are unaccustomed to heavy work, it will surely lead to sore muscles. If you trip or slip while lugging a heavy log, you could wind up with broken bones, which

Try rolling a log away from you along straight poles, as shown. The technique requires some bending but no lifting whatever. The poles elevate log bark above chain-dulling grit.

are easier to cope with if the log isn't lying on top of you.

Besides, such heavy work will tire you fast. You don't want to be tired if you have more chainsawing to do. Tired people have chainsaw accidents.

Accompanying illustrations show some means of lifting and moving logs.

You can also move logs quickly by tossing them end for end. Beginning as shown, keep your back straight and lift with your legs. A 200-pound log requires only a 100-pound lift at one end, and that end decreases in weight steadily toward zero pounds as your log approaches vertical. Then let it flop forward away from you. The disadvantage here is the tendency for the bark to pick up chain-dulling grit each time you let the log flop.

Dragging also lets you transfer some log weight to the ground. But missed steps can lead to falls, with the log landing on top of you.

The peavy with cant hook (left) lets you turn large logs. Log tongs (right) let partners drag large logs.

Scandinavian pry bars, shown in earlier chapters, have cant hooks that allow partners to sail logs overland without the bark abrading trousers.

Sequence continued next page.

What's easier? Lugging a log to a vehicle for later bucking to stove length? Or bucking to stove length where the tree falls and wheeling short logs to the vehicle? Terrain is the deciding factor. If you cut to stove length before hauling, you'll spend extra time in the woods all right, but you'll eliminate bucking and extra handling time at home. Then too, by cutting to stove lengths immediately, you can load your vehicle much tighter than full-length logs would allow. Besides, the wheelbarrow lets you avoid heavy lifting. Pneumatic tires are essential.

12
If You Decide to Climb

WHETHER TO CLIMB

Climbing is normally unique to residential tree work. Here limbs may need pruning or whole trees may need to be topped because there is insufficient felling room.

Climbing may involve easy work from a ladder as well as strenuous free-climbing and technical climbing that employs climbing spurs and rappelling gear. For the nonprofessional, two rules should be observed:

1. Don't climb if you can avoid it.
2. If you climb, don't take a chainsaw along.

Pros do use chainsaws in trees. But these guys are in top-notch physical condition; they use the smallest saws that can handle the tasks; and they know how to stabilize and secure themselves high up so that unexpected pushes and pulls from the saw won't result in lacerations or falls.

If you decide you must climb, remember that handsaws will usually be adequate for most aerial cutting tasks. Of course, handsaws make for more tedious cutting, but patience and perseverance are good old virtues to foster, especially when you are 60 feet up a tree. Up there, haste can lay you waste.

On the other hand, if the job is too big for handsaws, consider hiring a pro to do the work you're not equipped to handle. For example, you might ask for bids from various tree workers to top a tree, leaving the remaining bucking and felling to you.

Veteran tree workers enhance their simian climbing skills with climbing/rappelling ropes (right) that also serve as safety lanyards.

SAFETY BELT AND ROPE

Whether you stick to a ladder top or engage in free climbing, wear a safety belt with safety rope that can be fastenend to a sturdy tree limb.

The best professional belts feature a broad, heavy-duty waistband and either a sling that supports your rump or a loop for each thigh that cinches tight. Of the two belt options, the type with thigh straps provides better all-around support. If you fall a short distance wearing it, the safety rope will arrest you in a seated position. Both types of belt let you lean out from the tree to gain better sawing leverage. Using a handsaw, you can then gain the support and leverage afforded by one or both legs, your free arm, and the belt as well.

The next best choice is a nylon rock climber's harness. It's lightweight and close-fitting. It can serve as well as belts designed for tree work, and it costs only about half as much.

Next in effectiveness is the professional belt designed for telephone linemen and climbing construction workers. This consists of a broad waistband. It is less desirable than harnesses because, in a short fall, it can snap your spine. If it's loose fitting as well, it can crack your ribs. If you lean out against it for added sawing leverage, it presses against your kidneys.

The least effective of belts is probably the one most often used. It's a simple belt fashioned from rope. Of course, such a belt is usually better than nothing. But it will punish you as it arrests a short fall—perhaps snapping your spine, pinching your kidneys, crunching your ribs, and then scorching your armpits. After such treatment, you might wonder if you wouldn't have faired better falling all the way.

BOOTS FOR CLIMBING

Tree-climbing youngsters suffer no pain wearing tennis shoes. But kids are lightweight. The more you weigh, the more your feet will appreciate stiff-soled boots that evenly distribute the press of ladder rungs and the crush of branch crotches. In fact, if you stand in a steep branch crotch with a soft-soled shoe, you'll soon note a numbing sensation in your toes because the crotch will have pinched off blood flow. Boots are better than mere stiff-soled shoes because they ward off scraped ankles.

Sole exception: If you are merely pruning a valuable, thin-barked tree, stiff-soles may damage bark and thus make the tree vulnerable to insects and rot. In this case, consider using soft soles.

USING LADDERS

Next to your chainsaw, your ladder may be the most hazardous tool you own. Yet ladders can be indispensable aids in tree work. We'll confine our discussion to common straight ladders and extension ladders. Here step ladders are preposterous. The ideal (but costly and specialized) pruning ladders have side rails tapering to a point at the top and let you rest the ladder top in tree crotches, reducing the danger of the ladder's tipping sideward.

For most ladder work, you should employ at least two lengths of sturdy rope, in addition to the rope needed for the pulley of an extension ladder. The first rope is your safety rope (lanyard), which you should run from your safety belt to a sturdy limb whenever you stop to work. The second rope should be long enough to secure the ladder top to the tree so that the ladder cannot tip backward or twist away.

You may decide not to tie the ladder top to the tree if you plan to climb further into the tree and drop limbs that could strike and damage the ladder. In this case, you'll need a groundman to remove the ladder—the same groundman who should have been on hand to stabilize the ladder's feet as you climbed initially. More on the groundman later!

Aluminum or wood ladders do fine. If you are using an old wooden ladder, though, inspect it for defects. Check the rungs for rot and cracks as well as the sturdiness of rung-and-rail joints. On an extension ladder, check the operation of the rope and pulley, as well as the reliability of the rung-

For pruning from a ladder, tie the ladder top snugly to the trunk. The safety belt shown features a broad rump band and leather thigh straps that allow you to "sit" in space to gain added sawing leverage. Boy Scout books offer good knot-tying instructions and show how to whip rope ends.

locking mechanism. Laid into position against the tree, the foot of the ladder should be about one-fourth the ladder's extended length away (horizontally) from the point of support. (Avoid use of aluminum ladders near electrical wires.)

TREETOP TIPS

1. Secure your safety belt whenever you stop to work. Always secure it above you and pulled snugly enough that you would fall only a few inches rather than feet before the rope would arrest your fall.

2. Avoid cutting any limb above you or at your level if it could fall on you or jump its stump and hit you or your safety rope.

3. Use stub ends of cut limbs within reach to hang tools or extra rope. When hanging your handsaw, be sure its blade is beyond accidental reach—so that you can't possibly hit it should you slip and scramble.

4. Use a weighted throwing line or else a hooked bamboo pole to guide boom ropes through crotches that will serve as booms in lowering cut limbs to the ground. These aids save effort and help you avoid hazards of unnecessary climbing.

5. Use a relay rope to hoist tools to yourself rather than attempting to carry tools in one hand while you climb. Thus you'll have both hands free for climbing and securing yourself in the tree. With the aid of your groundman, you can raise and lower tools as necessary.

THE GROUNDMAN

A groundman is essential for all but simple pruning from a short ladder. A groundman can ensure your safety by stabilizing the ladder feet as you climb and work from the ladder. Otherwise, your leaning to either side can twist the ladder feet off the ground even though you may have lashed the ladder top to the tree.

Once you have left the ladder and begun free climbing of the limbs, the groundman can attach tools and rope to a relay rope you've carried up with you; whereupon you can hoist the tools to your perch. The groundman also handles the boom rope, lowering and releasing cut limbs.

For safety's sake, make sure the groundman wears a good hardhat and avoids working so close under you that falling tools or limbs could strike him. In fact, by employing a guide rope attached to the limb to be lowered—in addition to the boom rope—the groundman can guide the limb to rest at his feet without venturing under you.

It's also important that you and your groundman discuss your plans for each limb to be cut. You may want to lower some and let others free-fall. Before you start cutting a limb to be lowered, be certain that it won't be too heavy or awkward for the groundman to handle. Nor should a limb fall far before the rope arrests the fall; limbs that are no thicker than your arm at their stump can exert tremendous stress on a rope if allowed to fall far. Such a limb free-falling from a tall tree will impact the ground with great force, the stump end often driven deeply into the ground.

Placement of the boom rope is critical. Generally try to run it through a sturdy crotch somewhat above you, but far enough to the side so the cut limb swings away from you, pendulum fashion. Yet be sure the arc of the pendulum doesn't result in a return swing that carries the limb back, possibly hitting you.

ROPE SAVVY

Avoid using old, frayed ropes or those that may have been weakened by damp rot in basements. Especially avoid joining odd pieces of rope that neighbors may offer. Continuous lengths of sturdy rope designed for tree work are always the safest bet.

Size. Most tree workers use half-inch rope, whether of manila or synthetics. First, half-inch rope lays in the hand well, and can be coiled or thrown easily. Half-inch manila has roughly a 2,000-pound breaking strength. Half-inch synthetics have higher breaking strengths and may be more than twice as strong as manila in premium tree-work grades. In the past, many tree workers preferred manila because it resisted kinks and stretch better than synthetics. In fact, early synthetic ropes (and cheaper ones today) would stretch and recoil so much under sudden weight that they caused heavy limbs to "bounce" in the air, often endangering the climber. Thicker, stronger "bull" ropes are used for pulling trees over.

Manila vs. Synthetics. Aside from the higher strength of synthetic rope for a given thickness, synthetics absolutely won't mildew or rot as untreated manila can. Synthetics are lighter weight for a given thickness and length, which can be an advantage in longer lengths and for throwing rope ends over limbs and through crotches. Still, in manila's favor, it can be treated to resist moisture absorption, and it costs only about half as much as premium synthetics.

Whipping Rope Ends. Synthetic rope ends can be fused with a match flame to prevent fraying. Manila should be whipped with strong string. If you are in a hurry, you can give manila a temporary whip by knotting the end or wrapping it with tape. But knotted ends look terrible and can become cumbersome. Tape can fail if hot sun affects the adhesive or if sun and water affect the bond.

Rope Care. Abrasion from tree bark and from embedded grit can wear away individual strands and weaken the rope. To prolong rope life, avoid unnecessary bark abrasion and keep the rope out of dirt, especially a problem when ground is damp. Store manila in a dry place. Store synthetics away from direct rays of the sun.

CLIMBING SPURS

Many weekend tree workers are tempted to strap on spurs rather than borrow an extension ladder. But spur climbing is arduous work. It requires practice. And *it is extremely dangerous.*

Many weekend climbers use lineman's spurs, designed for climbing barkless power poles. Trouble is, these spurs have gaffs (spurs) that are only about two inches long. The gaffs on tree-climbing spurs are 3½ inches long so that they can penetrate thick bark and plunge into sapwood.

If a gaff should break loose from the bark, you might fall in staccato increments down the trunk, your safety rope banging your crotch into the bark until either you or the rope hangs up on something. Then too, if a gaff breaks loose, your leg thrust could drive the gaff into your opposite ankle. Aside from the discomfort this causes, it leaves you with bothersome decisions on extracting the gaff and then making a descent.

Best advice is to refrain from using spurs unless you are planning to turn pro. Even then, first take some lessons from a veteran tree worker.

II

Chain and Bar Savvy

Chain Maintenance. Of all saw maintenance tasks, none is as important as chain maintenance. A saw that runs poorly, or even unpredictably, can still cut wood as long as the chain is well maintained. But a well-maintained saw won't cut at all if its chain isn't up to the task.

Besides, a well-maintained chain cuts efficiently—saving you cutting time. It cuts smoothly—reducing wear on chain, bar, sprocket, and the saw itself, while greatly lessening kickback tendencies. A smooth-running chain also helps reduce vibrations that fatigue your hands and arms.

Many professional loggers who do their own sharpening don't get half the potential cutting efficiency and life from their chains. If the pros have trouble, no wonder so many nonprofessionals rely on sharpening shops. The fear of sharpening goes even further, though. Consider the fellow who can't wait a few days for a professional sharpening and so instead goes out and buys himself a brand new chain.

Alternatives? How about saws with built-in power sharpeners designed for special chain? Under ideal conditions, these sharpeners do a pretty good job. Yet the chain still needs occasional manual filing touch-ups and an awareness on your part of what a well-sharpened chain ought to look like. There are two additional drawbacks here. Mechanical problems in power sharpeners themselves often result in below par sharpening. And

This photo shows a full-speed chain cutting the end of an ash log. Typically, chain cutters rock back as they bite into wood, but only as far as the depth gauge (shark fin in front of cutter) allows. This rocking action raises the toe of the cutter off the guide bar rails and makes the heel of the cutter drag along the rails. For sharpening, these cutting dynamics, will help you appreciate the crucial relationship of cutter and depth-gauge heights.

these special chains lack the reduced-kick advantages of new-generation chains.

Upcoming pages explain how to get nearly 100 percent efficiency and maximum life from all types of chains by maintaining them yourself.

Ten Reasons for Sharpening Your Own Chain

1. If you understand a few simple concepts for the use of sharpening tools, you can maintain your chain as well as—or better than—the pros at the chainsaw shops.

Compare the old standard chain in back with a reduced-kick version by Oregon in front. On the old standard chain, the great gap between cutters and the sharply vertical fronts of depth gauges conspire to cause noncutting impacts against wood as the chain rounds the upper quadrant of the bar nose. These impacts momentarily hang the chain up, with the result that engine torque is violently transferred to the saw and bar and the operator's hands. This phenomenon, known as kickback, tends to whip the saw and bar upward in an arc and has resulted in about one-third of all chainsaw-related injuries. On the chain in the foreground, note the long, sharklike fin projecting from the tie strap directly ahead of each cutter. As the chain arcs around the bar, the tail of the tie strap rotates out slightly farther than on the flats of the bar, serving to deflect wood impacts from the depth gauge. Note that the depth gauge itself is sloped too, to ensure deflection.

2.   Sharpening is easy, requiring neither a high mechanical aptitude nor an engineering degree.

3.   Sharpening saves you money—more than the price of a new chain in just a few sharpenings.

4.   Sharpening is enjoyable. It lets you communicate more closely with your saw—while you are sharpening, while you are cutting, and whenever the cutting edges begin signaling the need for a touch-up.

5.   Sharpening saves you time you'd otherwise waste transporting your saw to a shop—especially if you'd have to wait around or, worse, come back later to pick it up.

6.   A chain getting heavy use should be touched up, at least, every couple of hours, or whenever sand, dirt, or grit has reduced cutting efficiency. And it's downright troublesome to deliver a chain to a shop for sharpening every few operating hours. Instead, many saw owners mistakenly continue cutting with a dull chain by applying more pressure to the bar. You can force a dull chain to scratch and scrape its way through wood a while, but this soon damages both chain and bar, sometimes beyond repair. Such damage also increases the danger of a broken-chain mishap.

7.   You can sharpen a chain as fast as shop technicians can. Your sharpening can be more precisely tuned to the type of cutting you anticipate.

8.   Most sharpening shops do a good job. Some do not. Some shop technicians—out of sloppiness or ignorance—don't modify their grinding techniques in relation to the types of chain they're sharpening or the specifications of the chain maker. Some tend to overheat cutters by grinding them too long, blueing the cutting edges, causing hard spots that dull files. Excessive grinding also removes excessive amounts of metal, shortening the chain's life. And a worn grinding disc, or wheel, can cause chain irregularities, as will a variety of incorrect settings.

9.   Most sharpening shops should carefully inspect your chain and advise you of trouble signs resulting from mechanical problems or your cutting techniques. Few shops go far enough here.

10.   A goodly percentage of saws awaiting professional sharpening have drive sprockets damaged enough to warrant replacement. Otherwise the sprockets will continue damaging chain parts at an accelerating rate. A sprocket costs only about 20 percent as much as a new chain and takes only a few minutes to replace. If you sharpen your own chain, you'll likely keep a closer check on sprocket wear than you would if you entrusted all to the shop.

There are some other reasons for do-it-yourself sharpening, but those above ought to serve for now.

CHAIN BASICS

Chain Parts. During the 1970s, chain parts from one manufacturer to the next looked pretty much alike. Yet concern over kickback injuries dictated that the industry find ways of reducing the incidence and force of kickbacks, which until then were accounting for about one-third of all chainsaw-related injuries.

With the coming of the 1980s, most manufacturers had begun offering chain designed to reduce kickback forces without loss of cutting efficiency. The changes occurred primarily in chain profiles—some only slight variations from the old standard, some dramatically different. (For a thorough discussion of kickback causes, forces, and preventative means, refer to the chapter on kickback.)

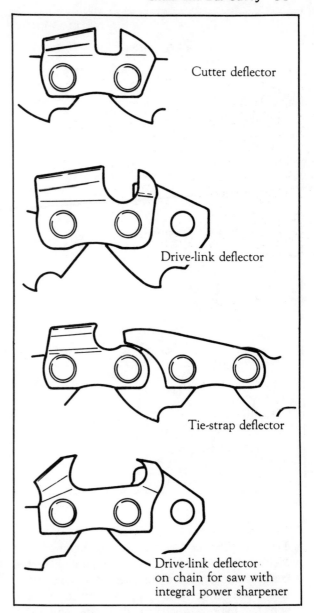

Cutter deflector

Drive-link deflector

Tie-strap deflector

Drive-link deflector on chain for saw with integral power sharpener

Ramps on both the tie strap and the drive link on this Sabre Tri-Raker chain by Townsend serve to broaden the facial area of the depth gauge ramp. These ramps greatly reduce potential kickback forces because they prevent the cutter's depth gauge from digging into wood as it arcs around the upper quadrant of the bar nose. (Photo by Townsend Chainsaw Co.)

To produce chain tending to yield milder kickback forces, manufacturers have designed an intriguing variety of chain profiles that help deflect wood from the front of the depth gauge as it rounds the upper quadrant of the bar nose.

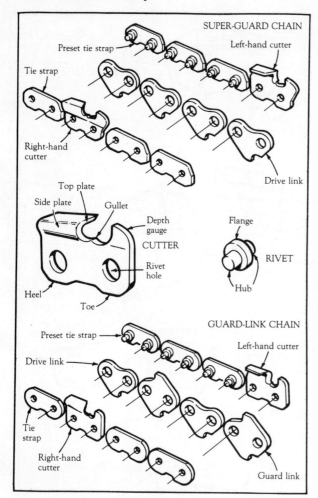

Exploded views of the two principal options in reduced-kick chain: The top view shows gently ramped drive links and cutters with deflector depth gauges. The lower view shows high deflectors on alternate drive links, and cutters with the old-standard depth gauge. (Adapted from Oregon Saw Chain drawing)

Left to right are "right-hand" cutters of chipper, semi-chisel, and chisel chain. Professional loggers usually prefer chisel chain for big timber when bark is essentially free of abrasives. The square edge of chisel chain results in a leading point on the cutter that lets chain cut faster, more aggressively, than chain with rounder corners. But the chisel point is highly vulnerable to sudden dulling from grit. So loggers who take on gritty bark usually prefer chipper chain or semi-chisel chain, which are less suddenly dulled by grit.

Types of Chain. Most chain makers characterize their chain primarily by the configuration of the cutter. Basically, there are three main categories: chisel, semi-chisel, and chipper—with slight variations and hybrids. Each has advantages and disadvantages when compared to the others, as discussed in the accompanying illustrations.

Another prime consideration is the frequency and spacing of cutters and noncutting tie straps. Here are the major distinctions:

• Standard chain is recommended for all general cutting tasks. It consists of one set of tie straps between each set of cutters and is designed for all-around use.

• Skip chain has two sets of tie straps between cutters. The extra space between cutters allows for clearance of chips for the saws used by pros in big softwood timber.

• Full-house chain has no tie straps. Here the close spacing of the cutters helps smoothen the cutting action on soft metals such as aluminum.

Makes and Sizes of Chain. Several saw companies manufacture their own chain. Others buy specially made chain with the saw company's own label on it. Other saw makers simply mount chain by various chain makers, thus giving each chain maker full credit or blame for the quality.

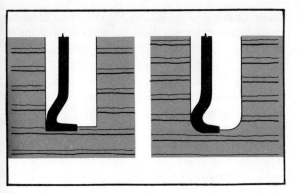

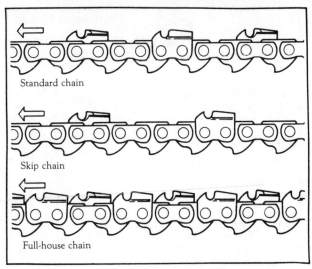

These are cross-sectional views of the kerfs made by chisel chain (left) and by chipper chain. Kerfs of both types of chain become wider as chain pitch (size) increases. Successive cutters are on alternate sides of chain, so each cutter cuts only about one-half of the kerf.

Frequency of noncutting tie straps reflects the type of cutting to be done. Standard chain: general cutting. Skip chain: lightweight softwoods such as cedar. Full-house chain: soft metals.

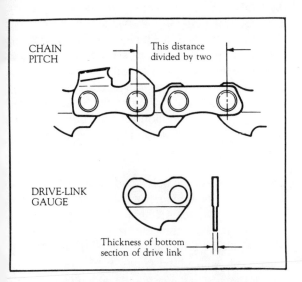

Chain pitch is the measure of half the distance between any three successive rivets. Smallest chains have ¼-inch pitch. Chains for big timber may be ½-inch pitch. When buying chain, be sure that its pitch matches the teeth on your saw's drive sprocket and that the chain's gauge matches the groove in your guide bar. Service shops carry charts with specifications. (Adapted from Oregon drawing)

Aside from cutter and linkage configurations, chain is labeled by the pitch of its drive links and cutters, and by the gauge of its drive tangs, as shown in accompanying drawings.

The tiniest pitch—¼-inch pitch for small saws—is .050 gauge, which is roughly the thickness of a dime. Chain pitches in increasing sizes from .325-, 3/8-, .404, and 1/2-inch are available in three gauges: .050, .058, and .063.

Saw makers specify which chain pitch and gauge, and which bar grooves and drive sprockets should be used with their saws. And these sizes are usually stamped into the metal of these components for easy identification. Although saw makers would prefer that you use only their own products, other makers of compatible components are alert to sales opportunities. Thus, dealers can select compatible chains, bars, and sprockets for you from charts giving specs on these items made by various makers. Chain code numbers are often stamped on the drive link.

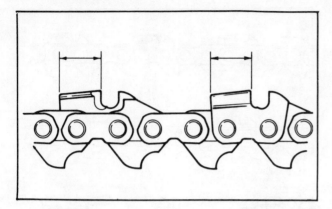

When sharpening, take pains to file or grind cutters back to equal lengths. This serves two purposes: (1) Top plates on cutters slope backward; if cutters are equal in length, they will automatically be of equal height and provide a reference from which to set depth gauges at uniform height. (2) Equal lengths of cutters also result in uniform distances between cutters and their depth gauges; this provides uniform amounts of rocking action as cutters bite into wood only so far as depth gauges allow. (Courtesy of Oregon)

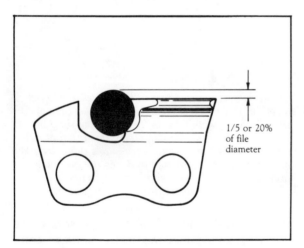

1/5 or 20% of file diameter

Depending on chain design, makers recommend that you sharpen with 10 to 20 percent of the file or grinder above the cutting edge of the top plate. On round-cornered chipper chain and on semi-chisel chain, this results in a nearly vertical profile of the side edge of the cutter. Most round-filed chisel chain should have a slight hook. The best mechanical sharpening guides automatically make your file or grinder protrude the desired amount above each cutter.

Chain Quality. Professional loggers and competition lumberjacks often swear by one brand of chain or another. And they may argue into the night which is best. Generally, the name brand chains are all good. Beware though of no-name chains, especially in discount centers. Chain makers with no name or reputation to protect may not ensure that the metallurgy, grinding, and inspection of each finished chain is top quality. Usually, you'll be safest to buy chain from a reputable saw dealer who will carefully match pitch and gauge to your saw and then check for sprocket and bar wear as well.

ESSENTIAL CONCEPTS

As Cutters Cut. On all cutters, it's the top plate that feeds the cutter into the wood. The cutter corner severs the wood fibers—thereby doing the hard work and taking the most punishment. This is especially true of square-cornered chisel chain, on which the leading edge is a point, rather than a curve. Conveniently in sharpening, if you hold the file in the correct position for the corner, the top and side plates will automatically become sharp too.

Correct Edges. Tools and techniques vary somewhat depending on whether you have square-cornered chisel chain, or semi-chisel, or rounded-cornered chipper chain. You can employ a round file or grinding rod on all three types of chain. Yet, with chisel chain, some pros instead use a beveled file for a difficult-to-file grooved, rather than concave, inner edge.

You'll be able to visualize appropriate sharpening angles best after studying accompanying illustrations and captions. *Caution:* With slight changes in chain design, makers of similar-looking chain may recommend slightly different sharpening angles as well as differing file or grinder diameters. It is no longer true that the very common 3/8-inch pitch chain should always be sharpened with a 7/32-inch file or grinder. Depending on the height of the cutters and the amount of wear on them, some makers recommend files of slightly greater or lesser diameters for 3/8-inch pitch.

Chain loops sold in factory packages are normally accompanied by sharpening instructions. On the other hand, chain dealers usually make up chain loops from 100-foot reels. In this case the dealer should be able to give you sharpening advice. Lacking that, you can attempt to match a file or grinder diameter to the factory-ground inner face on the cutter. At the same time you can determine filing angles by using either a protractor or the angle settings stamped into any of the simpler filing guides.

Usually, but not always, round-cornered chipper and semi-chisel chains need a 35 degree angle for general cutting, with the file or grinder held parallel to the top plate of the cutter. Usually, but not always, chisel chain gets a 30 degree cutting angle, with the file or grinder angled downward 10 degrees on the inside of the cutter.

Aside from filing angles, pay close attention to the amount that the file or grinder projects above the top plate. The less projection, the greater forward hook the cutter will have. The greater the projection, the more vertical or back-slanting the profile. On round-cornered chisel and on semi-chisel chain, the best profile is close to vertical or back-slanting to 85 degrees, rather than hooked. You can achieve this by allowing one-fifth of the file or grinder to project above the top plate. *Note:* A sharply hooked cutter will usually cut faster for a short time than a properly sharpened chain, but then wood impact will dull a hooked chain sooner because its leading edges are too acute, or thin.

These three series of drawings show filing angles that chain makers normally recommend for either their chipper and chisel chains and for both types of chain if used for ripping (lumber making). Note: The top plate angle of 30° or 35° combined with the 0° or 10° side slope will automatically give you the cutter profile angle shown. (Courtesy of Oregon)

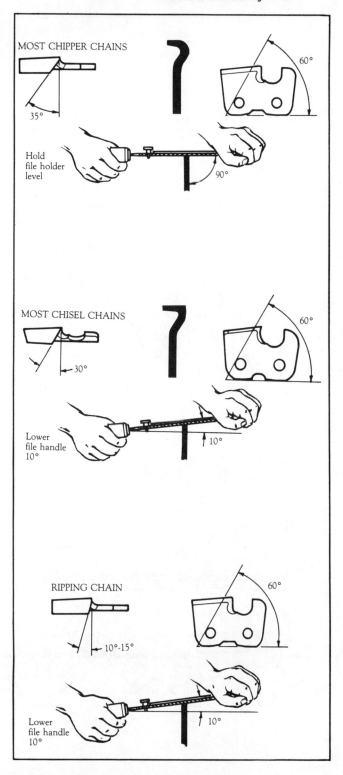

MOST CHIPPER CHAINS

35°

Hold file holder level

60°

90°

MOST CHISEL CHAINS

30°

Lower file handle 10°

60°

10°

RIPPING CHAIN

10°-15°

Lower file handle 10°

60°

10°

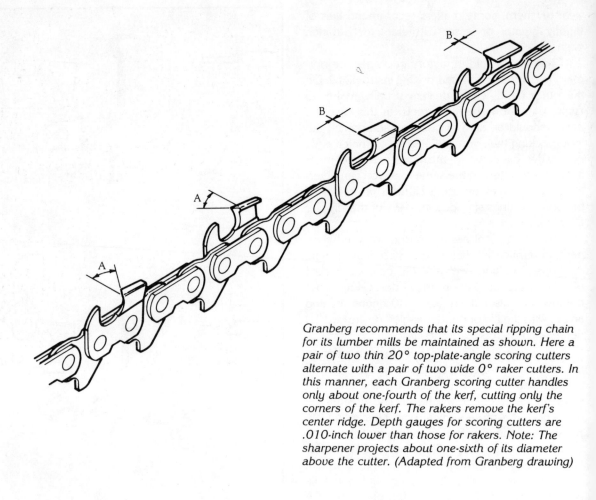

Granberg recommends that its special ripping chain for its lumber mills be maintained as shown. Here a pair of two thin 20° top-plate-angle scoring cutters alternate with a pair of two wide 0° raker cutters. In this manner, each Granberg scoring cutter handles only about one-fourth of the kerf, cutting only the corners of the kerf. The rakers remove the kerf's center ridge. Depth gauges for scoring cutters are .010-inch lower than those for rakers. Note: The sharpener projects about one-sixth of its diameter above the cutter. (Adapted from Granberg drawing)

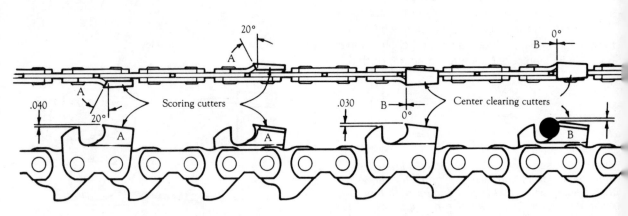

These two high-speed photos further illustrate the importance of depth-gauge height. In the first photo, tension of a cold chain causes the cutters to ride snugly along the bar rails—so snugly that depth gauges serve only to help clear loose wood chips from the kerf. The second photo shows the more normal dynamics of a working chain that has heated up and expanded enough to allow cutters to be lifted off the bar rails as the cutters bite into wood. Here the bite causes a rocking action of each cutter that is limited by the height of the depth gauge. Thus, cutters of a sharp chain will take uniform bites of wood that promote smooth cutting if depth gauges are filed to optimum and uniform depths below cutter heights.

Depth Gauge Settings. Cutters can be sharpened to perfection and still cut poorly, or not at all. Why? Each sharp cutter depends on the depth gauge (shark fin) just in front to limit the amount of bite the cutter takes. The accompanying high-speed photo shows how cutters actually rock back and lift off the bar only as far as the depth gauge allows them to. This is *one of the most essential concepts in sharpening.* In fact, one chain maker contends that incorrect depth-gauge settings cause more chain failures than any other factor.

If the depth gauges allow the cutter to take little or no bite, the chain will fail to cut and may be mistakenly thought dull. If the depth gauge allows too great a bite, the cutters rock back too much, causing excessive wear on the heel of each cutter as it impacts along the bar. Then too, uneven settings cause rough cutting, chain chatter, and excessive vibrations in the saw handles. For sure, depth-gauge setting is critical. We'll cover this in detail a bit later.

Field Sharpening vs. Home Sharpening. If you decide to do your own sharpening, you'll want to consider two quite different approaches. The first approach is to sharpen only at home and then begin each cutting day with two or three sharp chains. When the first chain shows signs of dullness, you can switch to the second and so on. Then if the last and final chain becomes dull, you can either quit for the day or field sharpen the chain.

The second approach is to field-sharpen a single chain as needed throughout the day. Field sharpening can be simple and can serve as a welcome break from cutting, provided you have appropriate tools. Yet, in cold weather, field sharpening can also be a brutal chore if your sharpening guide requires bare-fingered manipulation of multiple settings. Since some of the most precise sharpening rigs require quite a lot of bare-fingered fussing with settings, some loggers in winter leave them home or else perform their field sharpening inside a heated vehicle.

Which cutter is sharper? Yep! The right-hand one. The dull top-plate edge of the cutter at left is betrayed by the reflection of light. Usually a dulled chain will show more apparent abrasion along the outside corner than shown here.

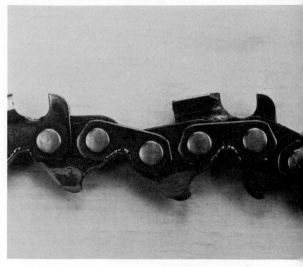

Peening and burrs in the notches of tie straps and cutters resulted from a worn spur sprocket. Peening here was so severe that many chain parts had tight points and all other parts moved stiffly. Note also the slight barbs on the drive tangs, which resulted from impacts with sprocket teeth. Sadly enough, as lengths of cutters evidence, this was a "young" chain that would have given much more service if the sprocket had been checked for wear before the new chain was installed.

When Does a Chain Need Sharpening? Owner's manuals prescribe a variety of ways for determining when a chain needs sharpening. Some suggest that you look closely for reflections of light off the leading edges on cutters, since sharp cutter edges reflect no light. Others suggest that you sharpen whenever your chain begins tossing out mainly sawdust rather than wood chips as it did when freshly sharpened. Both of these approaches have some merit all right, but as noted earlier, sharp cutters might still cut poorly and produce mainly sawdust if you fail to lower depth gauges sufficiently to allow cutters to take proper bites.

Provided you've set depth gauges properly, the surest sign of a dulling chain is its decreasing appetite for wood. A properly sharpened chain will cut hungrily downward into a log. It will need only the weight of the saw and bar to cut. If you must apply pressure to the bar to make the chain cut, stop cutting and either sharpen the chain or switch to sharp one. Otherwise, your forcing a dull chain will overheat it, fatigue chain parts, and cause excessive frictional wear on the bottom of the chain and along the bar rails. If the chain fails to cut straight, leading off to one side, stop cutting. Otherwise your bar and chain can be damaged beyond repair.

CHAIN TROUBLE DIAGNOSIS

The accompanying illustrations show common kinds of chain damage, suggest possible causes, and recommend remedies.

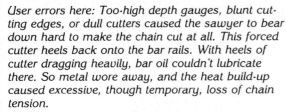

User errors here: Too-high depth gauges, blunt cutting edges, or dull cutters caused the sawyer to bear down hard to make the chain cut at all. This forced cutter heels back onto the bar rails. With heels of cutter dragging heavily, bar oil couldn't lubricate there. So metal wore away, and the heat build-up caused excessive, though temporary, loss of chain tension.

This cutter broke for several reasons. This chain, sharpened on a grinder, has a pronounced and deep hook in each cutter. Then, the chain failed to take adequate bites because high depth gauges prevented it. When even the hook failed, the owner bore down harder when cutting. This caused excessive friction between chain parts and the bar rails, evidenced by wear on the cutter heel. Deep filing of the gullet (between the cutter and the depth gauge) removed supporting sections of the cutters, tie straps, and drive links—weakening the cutters until sawing stresses broke the weakest cutter. Besides the waste here, broken chain ends could have injured the owner.

Saw Chain Troubleshooting Guide

Adapted from Oregon "Saw Chain
Trouble-shooting Guide."

**Most chain problems are
caused by 3 things:**

1. **INCORRECT FILING**
2. **LACK OF LUBRICATION**
3. **LOOSE CHAIN TENSION**

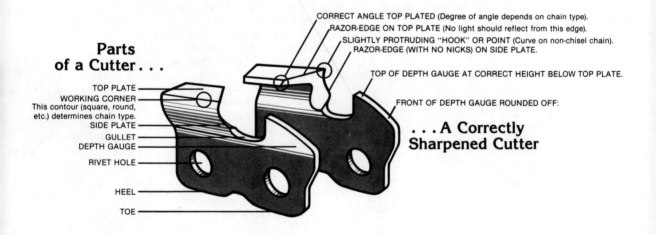

**Parts
of a Cutter...**

TOP PLATE
WORKING CORNER
This contour (square, round,
etc.) determines chain type.
SIDE PLATE
GULLET
DEPTH GAUGE
RIVET HOLE

HEEL

TOE

CORRECT ANGLE TOP PLATED (Degree of angle depends on chain type).
RAZOR-EDGE ON TOP PLATE (No light should reflect from this edge).
SLIGHTLY PROTRUDING "HOOK" OR POINT (Curve on non-chisel chain).
RAZOR-EDGE (WITH NO NICKS) ON SIDE PLATE.
TOP OF DEPTH GAUGE AT CORRECT HEIGHT BELOW TOP PLATE.
FRONT OF DEPTH GAUGE ROUNDED OFF:

**...A Correctly
Sharpened Cutter**

Here are some of the most common
signs to look for if your saw chain is not
performing right.

INCORRECT FILING

CONDITION: Backslope on side plate cutting edge. Cutter won't feed into wood.
CAUSE: File held too high.
REMEDY: Refile cutters to recommended angle.

CONDITION: Hook in side plate cutting edge. Cutters grab, cut rough.
CAUSE: File held too low, or file is too small.
REMEDY: Refile to recommended angle with right size file.

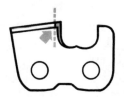

CONDITION: Flat top plate cutting angle. Chain won't feed into wood, won't cut.
CAUSE: File handle held too high.
REMEDY: Refile properly, at recommended angle.

CONDITION: Too thin top plate causes rapid dulling.
CAUSE: File handle held too low.
REMEDY: Refile properly at recommended angle.

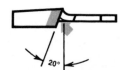

CONDITION: Top plate angle less than recommended. Causes slow cutting, excess wear on chain and bar.
CAUSE: File held at less than recommended angle.
REMEDY: Refile at correct angle.

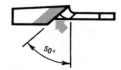

CONDITION: Top plate angle more than recommended. Side plate cutting edge is thin and dulls rapidly.
CAUSE: File held at more than recommended angle.
REMEDY: Refile at correct angle.

WEAR-CUTTERS AND TIE STRAPS

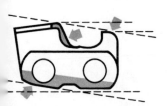

CONDITION: Excessive heel wear on cutters and tie straps.
CAUSES: 1. Blunt top plate filing.
2. Forcing dull chain to cut.
3. Forcing chain to cut frozen wood.
4. Low depth gauge settings.
5. Lack of lubrication.
REMEDY: Files cutters properly. Don't force dull chain to cut. Use oil freely.

CONDITION: Concave wear on bottom of cutters, connecting tie straps.
CAUSE: 1. Chain tension too tight.
2. Normal wear from undercutting (cutting with top of bar).
REMEDY: Adjust chain tension. Reduce cutting top of bar.

Sequence continued next page.

WEAR-CUTTERS AND TIE STRAPS *Continued*

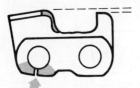

CONDITION: Cracks under rear rivet holes on cutters and opposing tie straps.
CAUSE: Excessive pressure on dull or misfiled cutters. Common during winter.
REMEDY: File chain correctly. Use oil freely.

CONDITION: Light damage on cutting edges of top and/or side plates.
CAUSE: Cutters hit sand or dirt, other foreign material.
REMEDY: File cutters to remove all damaged area.

CONDITION: Severe damage on either side of top and/or side plates.
CAUSE: Cutters hit abrasive material.
REMEDY: File cutters to remove all damage.

CONDITION: Peening on bottom of cutters and tie straps causes tight joints.
CAUSE: Loose chain tension. Result of dull cutters and forcing dull chain into wood.
REMEDY: Keep proper tension. Keep cutters sharp. Chain may need replacing.

CONDITION: Peening on front corner of cutters and intermediate tie straps. Causes tight joints.
CAUSE: Chain striking bar entry. Sprocket too small. Or loose chain tension.
REMEDY: Use proper bar and sprocket. Adjust chain tension correctly.

CONDITION: Excessive wear on bottom of cutters and tie straps.
CAUSE: Depth gauges too high. Cutting edge cannot get into wood.
REMEDY: Lower depth gauges to proper setting. Keep cutters filed correctly.

CONDITION: Edges burred and notch peened on tie strap.
CAUSE: Chain chatter due to loose chain tension and improper filing.
REMEDY: Correct chain tension. Refile chain properly. Replace sprocket if badly worn.

CONDITION: Peened notch in tie strap. Causes tight joints and broken drive links.
CAUSE: Chain run on badly worn spur sprocket or wrong pitch sprocket.
REMEDY: Replace worn sprocket. Chain may need replacing.

DEPTH GAUGES

CONDITION: Blunt depth gauge causes rough cutting.
CAUSE: Improperly filed depth gauge.
REMEDY: Round off front corner to maintain original shape.

CONDITION: Uneven depth gauge height. Chain won't cut straight.
CAUSE: Uneven filing.
REMEDY: Use correct depth gauge jointer to lower gauges evenly.

DRIVE LINKS

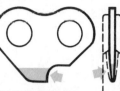

CONDITION:
Rounded
concave bottom.
CAUSE: Shallow
groove on bar tip.
REMEDY: Re-
groove bar tip. Bar
may need replacing.

CONDITION:
Nicked bottom or
back.
CAUSE: Cutting
with loose chain.
Or wrong pitch
sprocket.
REMEDY: Adjust
chain tension. Install
correct sprocket.
File off burrs.
Replace damaged
drive links.

CONDITION: Bat-
tered and broken
bottom.
CAUSE: Chain
jumped bar. Spur
sprocket hit drive
links.
REMEDY: Replace
damaged drive links,
sharpen tangs with
round file. Remove
burrs.

CONDITION: Sides
worn round at bot-
tom.
CAUSE: Chain
wobbled in bar
groove. Caused by
uneven cutters worn
bar rails.
REMEDY: Rework
bar rails and groove.
Correct chain filing.

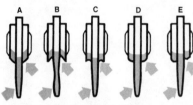

A. Open Bar Groove. **B.** Severe abrasion and wobbly chain on thin
bar rails. **C.** Rails not flat. **D.** Wobbly chain, rails too thick. **E.** One
rail too thin or soft.

CONDITION: Excess
wear on bottom of all
chain parts.
CAUSE: 1. Uneven filing,
worn bar rails
cause chain
to wobble.
    2. Excessive
pressure, try-
ing to make
chain cut.
REMEDY: File chain properly. Recondition
bar rails or replace bar.
Replace chain if
necessary.

CONDITION: Scars
on sides.
CAUSE: Loose
chain jumping off
bar.
REMEDY: Adjust
chain tension.
Replace bent drive
links.

CONDITION: Front
point turned up.
CAUSE: Drive links
bottoming in
sprocket. Sprocket
worn.
REMEDY: Replace
sprocket. Sharpen
tangs. Check for
burrs.

CONDITION: Front
or back peened.
CAUSE: Wrong
pitch sprocket or
prolonged chain
chatter.
REMEDY: Replace
sprocket. Adjust
chain tension. Chain
may be damaged
beyond repair.

Disc grinders with properly dressed wheels allow technicians to sharpen chain assembly-line fashion. In skilled hands, these machines perform precision sharpening. In the wrong hands, they alter hardness of cutters, waste cutter metal, and cause sawing problems. The disc must be dressed (shaped) properly; otherwise cutters will acquire incorrect angles.

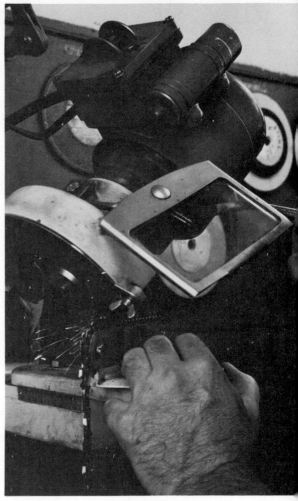

The Oregon Precision Filing Guide fastens over the bar and chain. Multiple settings allow highly precise filing at desired angles, heights, and depths. Once you've made settings, you simply take a few light filing strokes and then use your gloved hand to pull the chain forward to the next (in this photo) right-hand cuttter. Once you've done all the cutters on one side, you can set up to file the other side either by moving yourself to the other side of the bar or by staying put and swinging the saw and bar around. Similar guides are made by Granberg and Stihl. Cost: about that of a new chain.

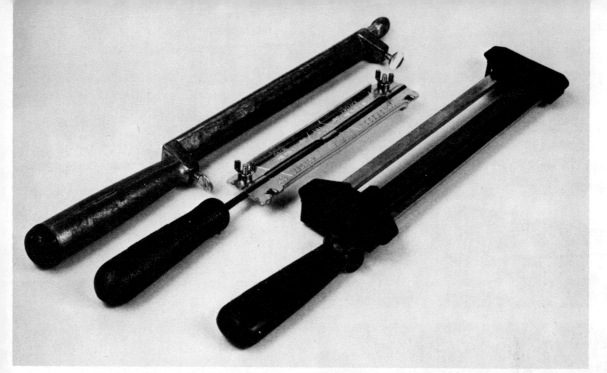

File guides such as these are designed to slide across the cutter top plate while a captive round file sharpens the cutter. The Sabre guide (left) uses only the cutter top plate of the affected cutter for support. The Oregon guide (middle) uses two points of support: the top plate and the depth gauge. The Pferd guide (right) uses two supports also: the top plates of the affected cutter and the cutter ahead. In addition, the Pferd guide incorporates a captive flat file that simultaneously keeps the depth gauge at .031-inch below the cutting edge. Of the three guides, the Oregon and the Pferd promote the more uniform filing, owing to the two points of support. But all three guides depend to a large degree on your "eye" and feel in achieving correct angles. And none tell you how far back you should file cutters. This is an important matter. Since cutter top plates slope backward, cutters filed back to varying degrees will have varying heights. Then since depth gauges are set in relation to cutter heights, a sharp chain can run "rough" owing to the basic inconsistency of cutter heights.

TYPES OF SHARPENERS

Based on the abrasive element and type power employed, there are four major types of sharpeners: (1) electric disc grinder, (2) hand-powered file guide, (3) electric hand grinder, (4) hand-powered fluted shaft.

Electric Disc Grinders. These begin in price at about $100 and cost upwards out of sight. Some are calibrated for the grinding of all types of chain. Others may grind either chisel chain or chipper chain. But, unless you're planning to run a logging crew or open a sharpening shop, you won't need a disc grinder. In fact, with a good, inexpensive file guide you'll be able to sharpen a chain as fast and precisely as you could with an expensive disc grinder.

File Guides. These are the most widely used and misused. And they vary widely in configuration and cost. Some of the best and worst cost only a few dollars. Other good rigs may cost as much as a new chain.

The popularity of file-mounted sharpeners undoubtedly owes to their low cost, to the availability of files of various diameters, and to the "feel" files allow the craftsman. Besides, they need no electrical power, so they can be taken afield. And they let you work at your own speed.

With the Carlton File-O-Plate (one side for left-hand cutters and depth gauges, the other side for right-hand), the only additional chain sharpening tools needed for precise chain maintenance are a round and a flat file. A bar clamp is useful but not absolutely necessary. The File-O-Plate works beautifully, but only on chain made by Carlton.

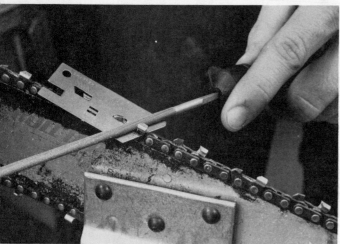

With your right hand on the front of the file and your left hand on the handle, file the left-hand cutter as shown, keeping the file angle parallel to the 35° rearward edge of the File-O-Plate. Use long strokes, and to avoid damaging file teeth, lift them clear on the backstroke. Then, remove the plate and check the depth gauge height (in the next photo). Pull the chain two cutters forward to the next left-hand cutter and repeat the procedure until all cutters on the left side are sharp and depth gauges are set properly. Then move to the other side of the bar to sharpen the right-hand cutters.

Check for the need to lower depth gauges, as shown, after every second or third touch-up sharpening. To do this, rest the midsection of the File-O-Plate over the cutter to determine if its depth gauge protrudes through the slot enough for trimming with the flat file.

In magazine ads, abrasive-rod electric grinders are portrayed as precision, no-effort sharpening marvels. But those that require mainly hand control can be difficult to use. Those with variable-speed motors are an asset because they allow you to approach the cutter with abrasive rod turning at low rpm. But high-rpm models can damage cutter edges severely with one false nick. Grinders that lock into preset grooves have greater potential for accuracy all right. But the best grinder guides have multiple settings as shown on the Oregon Precision Filing Guide earlier. Hand grinders cost twice as much, are less portable, and require just as much effort to use as hand-powered file guides.

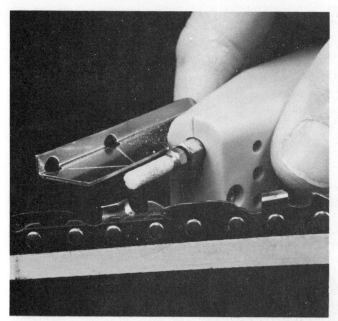

Electric Hand Grinders. These have abrasive rods that mount into chucks or rotating shafts, which may be linked to fixed, high-rpm motors or to variable-speed electric drills. Grinding rigs that offer multiple stationary settings help promote precise sharpening. But these cost about twice as much as a new chain. Less expensive grinders require that you depend on your eye and hands almost entirely. These tend to cause imprecise grinding because the eye is fallible and because the fast spinning of the abrasive rod makes it unruly.

The appeal of electric hand grinders may be the assumed accuracy and convenience the consumer attaches to electrical tools. But in some electric sharpeners at least, the assumption is questionable. Sales literature sometimes brags about the versatility of grinders that can be powered by flashlight batteries, 12-volt batteries, or household AC. Yet, by contrast, file-mounted sharpeners need no electrical umbilicals whatever—an important advantage, especially for field sharpening.

Fluted-shaft Grinders. At least one company offers a fluted carbide shaft that allows you to sharpen round-cornered chisel chain or semi-chisel chain at fixed, nonadjustable angles. The fixed angles don't allow you to alter settings at all. Yet, sharpening with this rig can be easy and fairly precise for general cutting, if you can envision how the sharpened chain should ultimately look while you make basic settings.

This sharpener by Gamn features a fluted carbide shaft that you turn with a hand crank. Its sharpening angles are fixed at about 33° for the top plate and 0° slope, so it can be used only on round-cornered chipper chain and semi-chisel chain. Cutting action is smooth and can be precise if you mount the unit at the correct height. Shafts and housings are available for chains of common pitches. Caution: Because the shaft devours chain hungrily, you might remove more metal than you want to, when making a tentative test sharpening.

Most files are imprinted with their diameters; some are not. Here files of 5/32, 3/16, and 7/32-inch protrude from the file-measuring section of the plastic Pitch-N-Gauge by Granberg. Note: if your chain gives you long life, some makers recommend that you use a smaller-diameter file after you've filed cutters about halfway back. Follow instructions of the chain maker here.

SHARPENING BASICS

If You Use a Round File. Be sure that your round files have a uniform diameter for the entire length. Some people mistakenly use a tapered "rat-tail" file on chain cutters. This causes highly irregular cutting edges.

Also be sure to use the file diameter prescribed by the chain maker. If your chain came without instructions on file diameter, bring your chain to a dealer and test-fit a few of the dealer's file sizes to determine which most closely fits into the factory-ground curl of the cutter. For example, most 3/8-inch pitch chain for large saws is ground for files of 7/32-inch diameter. But 1/4-inch and 3/8-inch pitch chains for small saws are designed to be filed with 5/32-inch files. Of course, you can resharpen a chain using a slightly larger or smaller round file than prescribed and still obtain good cutting efficiencies from your chain—though probably not top efficiencies. In fact, some chain makers recommend that you use a slightly smaller file after you've filed back about half of the original metal on the cutter. This is because the top plate of the cutter slopes backward, and thus the file used on a

new cutter might eventually begin cutting into drive links on a worn-back chain if you continued to file with the original file diameter.

On the Rockwell C scale of hardness most cutters are rated at about 53 to 55 RC. This allows for the best edge retension and durability. Files, themselves, are usually rated at about 62 on the Rockwell C scale and so can effectively shave chain metal away. Were the chain itself rated at a hardness over 62, it would undoubtedly be too brittle to survive the extreme stresses of wood cutting. Sometimes, though, excessive disc grinding of chain will cause the chain cutters to overheat and become blued. Blued cutters may be so hard that they will dull a file before you've sharpened a few cutters. And you may wear out more files in subsequent sharpenings until you've removed the brittle, blued metal and gotten back to the unblued metal. The hardened, blued cutters, can best be salvaged by *light* grinding with a grinder.

The best files carry imprints of their diameter and maker. No particular grid or cutting face on files seems to be clearly superior to others. Again, the main concerns are correct file diameter and sufficient remaining abrasive face on the file.

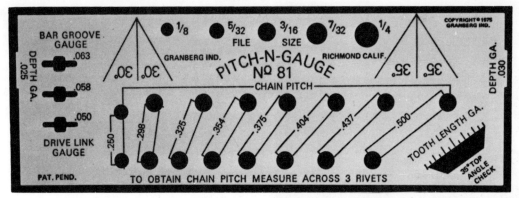

Just over five inches long, the pocket Pitch-N-Gauge by Granberg lets you measure files and grinding rods, chain pitch, drive-tang gauges, cutter-tooth lengths, depth gauge settings, and top-plate angles. It's handy in the tool chest and with you in the car if you spot a used saw at a tag sale.

If You Use an Electrical Hand Grinder. As with round files, grinder diameters are crucial. Precisely manufactured grinders will rotate on the axis established by the chuck of the electric tool you use. Thus it's important that the grinder shaft be precisely on center and that the grinder itself be perfectly round.

Rod grinders can be used with motor drives ranging from the 400 rpm slow speed of a variable-power electric drill up to the 30,000 rpm of high-speed drills or grinding tools. But the low rpm of electric drills makes for slow tedious grinding.

As with disc grinders, prolonged grinding on a single cutter can overheat the cutter enough to blue it—giving it hard spots. So if you must grind a damaged cutter back considerably, bounce the grinding rod lightly against the cutter to avoid burning it. All cutters should be ground back to uniform length, so if you must grind a chain extensively, grind each cutter lightly, moving to successive cutters until you've moved the entire chain around the bar as many times as necessary to grind back all cutters uniformly.

You can also apply a special waxy or soapy compound to grinders to minimize heat build-up and minimize the amount of metal dust the grinder face absorbs.

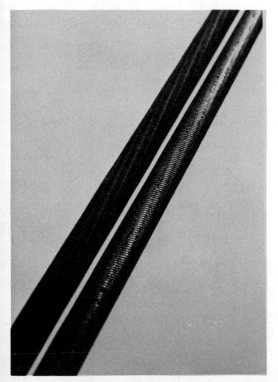

Cutting edges on new and worn files may feel about the same to the touch. But reflected light gives away the worn file, if unaggressive abrasive action hasn't already. If you use your saw often, plan to touch-up sharpen oftener. Remove all oil from cutters before filing. Then to prolong life of files, apply only light strokes. Individual files are relatively expensive at most dealers, but files come at attractive unit prices in boxes of 12 from some mail-order dealers.

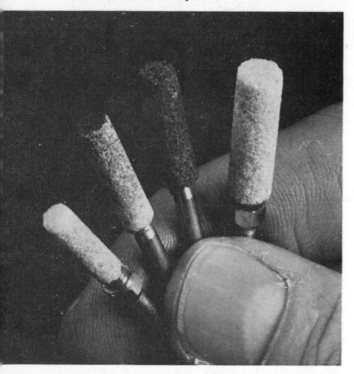

Grinding rods (stones) come in all standard chain sharpening diameters and fit chucks of portable electric drills as well as those of electric grinders. The rods tend to cost more than round files and give fewer sharpenings. If you apply a special waxy compound to them, the rods will cut cooler and absorb fewer metal fragments that otherwise reduce desired abrasive action.

This screw-tightened chain vise ("T bar") in the Oregon Precision Filing Guide prevents the chain from tilting in the bar groove as you file diagonally across the cutter. Lacking such a vise, you need to tighten chain tension and hold the chain in place with your free hand.

Oregon's bar vise has a prong on each side that can be driven into stump ends or horizontal logs.

With this stump vise you jam the side plates into a saw kerf before tightening the plates onto the guide bar.

For the stump vise in the preceding photo, prepare a 5/16-inch kerf. Do not attempt a boring cut—a dangerous and hard-to-manage cut in resistant grain of a log end. Instead slice with the bottom of the bar from the far side of the stump inward.

The handy Saw Jaw vise screws into a saw kerf, either in a stump as shown or along the topside of a downed log. When mounted on a downed log, it pivots 180° for left-hand and right-hand sharpening. Saw Jaw is available through saw dealers or Westwood Ventures Corp., 6018 Highway 179, Sedona, AZ 86336.

Small, home bench vises serve well for sharpening if they can be swiveled horizontally so that you can address first one side of the guide bar and then the other, or if they let you tighten up so you can address both sides of the bar. The jaws of the vise shown were too shallow to allow the chain to be pulled through. So wooden elevator plates were inserted on both sides of the bar to allow necessary clearance.

Here are four sparkplug gapping wires lain over a dime to help you visualize depth-gauge relationships to cutter heights. The wires (left to right) are .025, .030, .035, and .040 inch thick. The dime is about .050 inch thick. Most new chains come with depth gauges set at .025 to .040 inch below the height of cutters. If you lower depth gauges on new chain, the cutters will tend to bite too deeply into wood. This causes rough cutting and chain chatter—and fatigues chain parts.

Vises for Sharpening. Vises can serve in two important ways: (1) to keep the chain itself from tilting or shifting laterally on the bar rails, (2) to keep the saw bar steady. Many sharpening guides have chain vises built in. Some of these are shown in accompanying illustrations.

Normally, you won't need a bar vise when sharpening by means of an electric hand grinder or a fluted-shaft grinder. These grinders require only light pressure to the cutters in the direction of the saw engine, and the engine stabilizes things. A vise for the bar itself is a great asset when you use a file though. This is because the diagonal stroke of the file tends to push the bar away from you, unless you grip the bar with your free hand.

Special field vises are designed for insertion into kerfs in log ends or for driving into logs. Small home bench vises work well too. If you use a bench vise for sharpening *with a file*, the vise should swivel horizontally. Such a vise lets you rotate the saw and bar's left and then right sides in front of you.

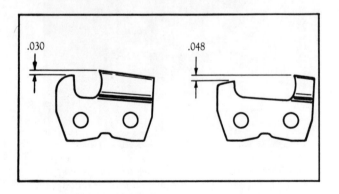

.030   .048

An old filed-back cutter won't take the same size bite of wood as when it was new unless you lower depth gauges slightly as you file cutters back. The Carlton File-O-Plate ensures that you lower depth gauges on Carlton chain gradually in relation to cutter height, as shown in this illustration by Carlton.

How to File Depth Gauges. Many chain failures result either from incorrect filing of depth gauges or neglect of them. The accompanying illustrations show details. As you study them, keep these thoughts in mind:

• For new chain of from 1/4-inch to 1/2-inch pitch (that's mini saw chain up to chain for big-timber saws), manufacturers usually recommend that depth gauges be only .025 to .040 inch below the top of the cutter. To visualize these depths, figure that a dime is about .050-inch thick. So depth settings on small chains need be only about half the thickness of a dime. Some chain makers stamp recommended depth settings into the depth gauge itself.

• As you file cutter back appreciably, you can increase the depth setting slightly, as illustrated,

Most hand-held depth-gauge filing guides rest on cutter top plates. A depressed groove in the guide allows excess metal on the depth gauge to protrude for filing. Since cutter top plates slope backward, all cutters must first be filed to uniform lengths so that depth gauges can be set at uniform height. With this type guide, depth settings cannot be increased as the cutter is filed back, and shifting of the guide can slightly dull newly sharpened cutters. After lowering depth gauges, round off the front corners as they were when new. (Courtesy of Oregon)

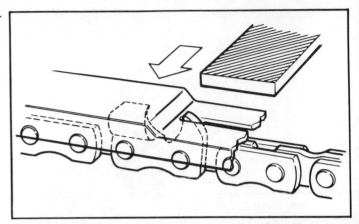

because the increasing distance between each cutter and its depth gauge and the rocking action of the cutter wouldn't allow the cutter to take as complete a bite as when it was new. Thus, the cutter won't take as full a bite unless depth gauges permit a full bite. For example, for a cutter that originally had a depth setting of .030 inch, you would need to drop the depth to about .048 by the time you've filed the cutter way back. (Oregon recommends that you maintain one depth-gauge setting on their chain throughout the life of the chain.)

• Many makers of chain and sharpeners advise you to lower depth gauges *before* sharpening the cutters. They advise this because many depth-gauge guides must be lain on top of cutters and so tend to dull cutter edges if you press down hard when filing the depth gauge. But such advice causes problems whenever you need to file cutters back significantly. This is so because cutter top plates slope backward, and you can't truly file depth gauges to a uniform depth along the chain until you've finished filing the cutters. So if the cutters require significant filing, lower the depth gauges after filing.

Best advice is a compromise between manufacturer advice here and your own horse sense. First file all cutters uniformly. Then lower all depth gauges uniformly. Then lightly touch up all cutting edges again. If you are willing to go this far, though, you are close to becoming a sharpening fine craftsman—if not a fanatic.

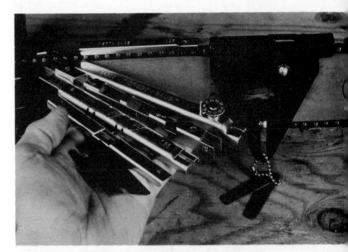

This array of depth-gauge guides deserves a closer look. In the palm of the hand (left to right) are the following: (1) Carlton File-O-Plate which automatically guides you in lowering depth-gauge settings as chain is filed back; but this guide works only on chain manufactured by Carlton, though much Carlton chain is sold without the Carlton imprint by a number of major saw makers. (2,3 and 4) Shown next are Oregon, Stihl, and Sabre fixed depth guides. (5) The fifth gauge is a no-longer manufactured Oregon Gaugit with turn-style calibrated depth setting, which you might still find at a saw dealership or at a tag sale. Mounted on the saw bar, left to right, are an Oregon end-depressed fixed guide and the gauge sold by Pioneer that allows precise adjustment of a screwmounted guide mechanism. Of all the guides shown, the Pioneer allows the most precise adjustments of settings on all makes of chain. Note: For all guides shown, uniformity in lengths of cutters will determine ultimate uniformity of depth-gauge heights.

The safety-edge file (left) has rounded sides that lessen the chance of your accidentally nicking and dulling a cutter while filing depth gauges. The all-purpose flat file (middle) is commonly found in hardware stores. The bevel-edge file (right) is sometimes used by pros for filing a grooved cutting edge into chisel chain.

If you need to restore the sawdust-raking hook of the drive tang, also remedy the source of wear. For this, consult the chain trouble diagnosis chart earlier in this section. (Adapted from Oregon drawing)

After sharpening chain, rinse off filings in kerosene and then soak the chain in clean motor oil until the next use. This prevents corrosion and ensures instant lubrication of chain and bar the next time around.

Restoring Hooks on Drive Tangs. You probably won't need to file the front of the drive tang, provided you've maintained the bar and the rest of the chain properly, and provided you've kept a worn drive sprocket from damaging the toe of the drive tang. If your drive tangs are damaged or worn, consult the chain trouble diagnosis chart earlier. You must first remedy the cause of damage and then restore the forward hook on each tang so that it can remove wood chips from the bar groove effectively. A round file will do the trick.

Cleaning the Bar and Chain. Most nonprofessionals conclude the cutting day by slipping a bar safety scabbard over the chain and bar and then storing the saw until the next use, which may be weeks or months away. But this is bad practice.

Chain should be stored in an oil bath (clean motor oil will do) whether for a day or many months. The oil serves two main purposes. First, it prevents corrosion. Second, it penetrates chain joints and lubricates mating surfaces. This ensures that the chain has immediate lubrication during the next cutting session before bar oil from the saw's reservoir begins to reach all chain components.

The bar too should be cleaned and oiled. This is primarily to remove the mixture of oil and sawdust that blocks the chain-oil inlet holes and builds up inside the bar groove. If this mix is allowed to harden, it becomes difficult to scrape out completely.

Sharpening should precede cleaning and storage, rather then follow. Here's why: After removal from an oil bath, a chain is so heavily oiled that sharpening becomes a messy proposition, and oil on the file will prevent its teeth from cutting. Besides, sharpening leaves metal filings and grit in the chain's joints. You should clean these abrasives from chain and bar immediately after sharpening, and before the oil bath. The simplest way is to coil the chain into a half-inch bath of kerosene in the bottom of a pie tin or coffee can and then slosh the chain about. Then lift the chain out to inspect for trouble signs before scraping off caked-on gunk and then storing the chain in a bath of motor oil. Here a lidded plastic tub or coffee can keeps both oil and chain clean.

Sawdust and chain oil become a glutinous mix inside bar grooves and bar oiler holes. If allowed to dry there, the mix hardens and, during the next cutting session, inhibits free flow of bar oil. Best bet is to clean bar grooves and oiler holes after each cutting session. Then oil the bar and wrap it with heavy paper to keep the bar dust-free and to inhibit corrosion.

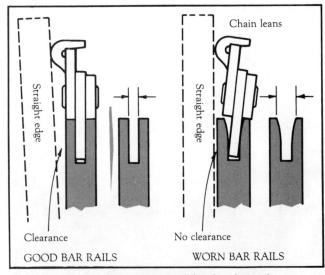

Worn or too-wide bar grooves will let the chain tilt as shown when you supply a straightedge. With chain removed from a worn bar, file off any burrs that may have built up along bar rails. If the bar has been strongly pinched in a kerf, also survey it for bends. To prevent wear primarily on one rail, rotate the bar (top for bottom) before beginning each day's work. To prevent general bar wear, maintain the chain properly, use plenty of bar oil, and don't apply force to make chain cut. (Courtesy of Oregon)

The finger points to the chain-tensioning nub on the nut that floats along the chain-tensioning screw. Clockwise turns of the screwdriver move the nub to the right, forcing the bar away from the sprocket. On some saws, this screw and nub are part of the routinely removable clutch housing assembly, rather than the engine case as shown here.

To clean and maintain the bar, scrape out all gunk from the bar groove and oil inlet holes. Then inspect the bar rails for wear. The rails should be square. If not, have them squared. If you find burrs, remove them with a flat file. (Accompanying illustrations show kinds of abnormal bar wear.) Then before storing the bar, give it a light coat of motor oil and wrap it in plastic or newspaper. *Note:* On a sprocket-nose bar, also check the nose and sprocket for adequate lubrication and damage. The sprocket itself is designed to lift the chain off the bar rails as it rounds the nose. When the sprocket becomes so worn that chain touches bar rails, it's time to replace the sprocket-nose assembly, if replaceable, provided the body of the bar is still in good condition. Unless you have a good bench grinder and belt sander, leave installation and grinding of the newly mounted sprocket nose to a chainsaw dealer. Otherwise, hand filing to make bar and sprocket rails mate is a difficult and tedious task.

The same nub shown in the previous photo here projects into the mating hole in the bar. Thus, with chain and bar mounted, turning the chain tensioning screw moves the bar, tightening or loosening chain tension. Seems simple: But many saw owners forget to make sure that the nub and the bar hole have mated before tightening the nuts over the two bar studs shown. This results in an unstable bar and a chain that loses tension rapidly. Further damage can occur if the user attempts to turn the chain tensioning screw when the two bar nuts are fastened down. Then, turning the screw can strip its threads, corkscrew it, or force the floating nut into the casing. It happens!

Here is a comparatively light-duty bar mounting assembly on a Wen electric saw. The screw tensioning nub mates with the bar hole just below the single large bar mounting nut.

SETTING UP FOR CUTTING

Remounting Bar and Chain. Until you get some practice, you may wish you had three hands for this task. At first, you'll swear that it's not as easy as your owner's manual indicates. Yet once you've become accustomed to the mounting assembly on your own saw, you'll wonder why you had apprehensions at first.

The trick is to place the saw itself on a stable surface at a comfortable working height. If you must instead place the saw on the ground, you'll need to drop to one knee and be especially careful not to let the chain or bar touch dirt. (Abrasives on log bark are dulling enough. So avoid coating cutters with abrasives before you even start cutting.) A home workbench is ideal for mounting chain. But in the field you can also use a flat stump. Yet, after you've practiced, you'll be able to mount bar and chain while seated on a felled long, leaning the saw engine against you as you work.

Since bar-mounting assemblies are slightly dif-

ferent from saw to saw, follow your owner's manual. Still, the accompanying illustrations may give you some useful tips.

Adjusting Chain Tension. If your bar has a sprocket nose, you should adjust chain tension so that cutters and tie straps on the bottom side of the bar touch the bar rails. Yet, tension should still be light enough to allow you to pull the chain easily around the bar. If your bar has a hard nose, adjust chain tension so that cutters and tie straps on the bottom side of the bar nearly contact the bar rails. Hold the nose of the bar up while adjusting and tightening bar nuts.

Brand-new chain should be soaked in motor oil a few hours before mounting. Then it needs a few minutes turning freely at half-throttle and then in light-duty cutting before mating surfaces are smoothed and chain rivets are properly seated. Be sure to oil heavily during chain break-in. A new chain will loosen up significantly before it is fully broken in. Watch for this apparent loss of tension and then retension accordingly.

Hold the nose of the bar up while tightening bar nuts—this after having held the nose up while tensioning chain by means of the chain tensioning screw.

For sprocket-nose bars, manufacturers all recommend that you tension chain so that cutters and tie straps on the bottom of the bar just touch the rails, but with chain only tight enough to allow your pulling it around the bar easily. Because there is more chain and bar friction around the nose of hard-nose bars, makers usually recommend that there be enough space between cutters and tie straps in relation to bar rails to allow insertion of a dime, as shown. In either case, hold the tip of the bar up while adjusting the bar and when tightening bar mounting nuts.

If you've tensioned a chain properly, it will lose some tension as wood friction and metal friction heat it, causing it to expand. Insufficient bar oil leads to metal friction that may burn the metal and cause it to smoke. Yet, even if the chain gets enough lubrication, wood friction from steady cutting on a large log can also cause the chain to heat up, expand, and sag. When this occurs, let the chain cool a bit before proceeding. As the chain cools, you may hear the unmistakable ticking sound of contracting metal. Watch as the sagging chain pulls up tight again against the bottom of the bar. *Note:* When you saw at full throttle, oil may not always get all the way around the bar. Thus it's wise to ease up now and then to let oil penetrate abrading surfaces better so that it isn't thrown off before it is carried around the bar.

Caution: To ensure adequate bar/chain lubrication in subfreezing weather, dilute your bar oil with kerosene or else use specially formulated oil.

Tom Bildeaux, world-class competitive lumberjack, here uses a top pull to tension chain on 24-inch bar at Homelite's Tournament of Kings, where he has been a perennial top contender. Many competitive sawyers set extremely slack tension for speed cutting events in order to promote frictionless chain travel. For woods work though, these same experts tension chain considerably more, in the interest of safety and chain life.

CHAIN REPAIR AND SPROCKET REPLACEMENT

Breaking and Mending Chain. You'll probably never need to break a chain (remove a rivet) unless you take up logging or chain-sharpening as a pro. Nor would you need to know how to mend a chain, that is, install chain components and then peen new rivets. Here's why.

If you become a pro, you can save money on chain by buying it in 100-foot reels from specialized dealers. In this case, breaking and mending become essential skills. But if your woodcutting doesn't wear out more than a few chains each year, you'll be well off simply to buy chain loops made to fit your bar (either through dealers or mail-order houses).

Chain parts need to be broken out (1) if you want to replace damaged parts or (2) if you want to shorten or lengthen a chain for mounting onto a different bar. Then, whenever you install new parts, you must shape them to match like com-

ponents on the original chain. Otherwise, the new chain parts will cause the chain to cut unevenly and precipitate damage to the bar and the old chain.

Besides, a chainsaw dealer has tools that let him break and mend in the wink of an eye. His most time-consuming tasks include sorting through trays for replacement parts and then dressing the new parts to match the old ones.

Tools for Breaking and Mending. You can break and mend chain using an inexpensive anvil and punch. The anvil has grooves that accept chain of various sizes. Then with hammer blows to the punch, you can punch out a rivet. But before replacing chain parts, shape them as near as possible to like components on the chain. Using the anvil you then must carefully peen the rivet over. This requires painstaking skill. Excessive or misdirected blows of a peen hammer can mushroom rivet heads excessively, resulting in a tight joint. When you are sawing, one tight joint can cause a link to break somewhere between the sprocket and the tight joint.

Costing little, the punch and anvil lets you punch out rivets and then peen new rivets into place. Slots in the anvil accept chain of all sizes. Caution: Breaking and mending procedures require forceful hammer blows that, if misdirected, could cause chain joints to bind.

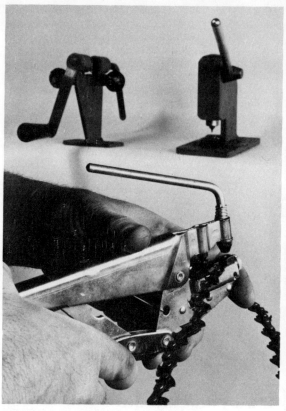

In this ingenious tool, Granberg has added chain-breaking and rivet-spinning capabilities to a trusty Vise Grip. The Break-N-Mend tool costs only about as much as a good chain. Shown in background are a bench chain breaker and a bench chain spinner, used mainly in pro shops.

Of course, if you use an anvil and punch, be sure to follow the manufacturer's instructions to a T. Remember, besides the tools, you'll need spare chain parts, which you may need to obtain from a dealer. He may have to bill you as much for chain parts, which you might never need, as he would bill you to break and mend. Plus the dealer would use far more precise tools than your anvil and punch.

Easier-to-operate and more precise break-and-mend tools range in price from the cost of a new chain to that of several chains. Unless you'd plan to use break-and-mend tools frequently, the payback on them might not occur for many years.

When Are Chain Repairs Worthwhile? Chain may be worth repairs if you damage a few cutters on a hidden nail. But if the chain simply breaks on you under a cutting load, *beware!* It broke at a weak point, and there may be other weak points resulting from long-term abrasion of rivets and seats, or from previous cutting impacts, or from a worn sprocket. Check the nature of the break and then inspect the whole chain for trouble by comparing it to the chain trouble diagnosis chart, earlier in this book. Remember, it's false economy to purchase repairs for a chain that will soon break elsewhere. Besides, the next break could allow chain ends to whip your way.

Left to right are (1) an unused seven-tooth spur sprocket fused to its clutch drum, (2) a used sprocket showing tooth damage common to most saws awaiting sharpening at service shops, (3) an astoundingly damaged sprocket. In the third case, the owner complained that the saw was no good because whenever he attempted to cut, the chain wouldn't turn. Of course, the deeply worn groove in the sprocket let the sprocket turn without engaging the chain's drive links. Note: There's also a deep groove worn around the clutch drum by a chain brake that must have been engaged often with the chain turning.

Changing Drive Sprockets. As a general rule, one sprocket's life is usually about equal to the life of nearly two chains. This is one of several reasons why you should consider using two chains alternately. Thus, by the time the two chains are worn out, the sprocket is too.

The two-chain, one-sprocket rule is a useful guideline. Yet a damaged sprocket can ruin chain quickly, and vice-versa. Chain damage can result from insufficient lubrication, excessive chain tension, or improper filing. Also, if a loose chain should fly off the bar and rattle on through the sprocket housing until the engine bogs down, the result may be damaged chain or sprocket, or both. Or simple sprocket wear will inflict further damage as chain and sprocket attempt to mesh. Plus, a worn sprocket can cause burrs on the notch of the chain tie straps and cutters. These burrs can tighten joints all around.

If you inspect the chainsaws awaiting sharpening at a saw shop, you'll find that most sprockets are damaged badly enough to warrant replacement in order to prolong the life of the chain. A sprocket costs only about 20 percent as much as a chain, so there are economies there. Some dealers will replace sprockets as needed, for the good of the chain. Others won't. One philosophy sells more sprockets. The other sells more chain. *Note:* Some dealers may be reluctant to change a sprocket for fear that you'll feel you were billed for unnecessary parts and labor. But dealers could alleviate that concern by attaching the damaged sprocket to the repair ticket as evidence. A full-service dealer will also spell out the whys and wherefores.

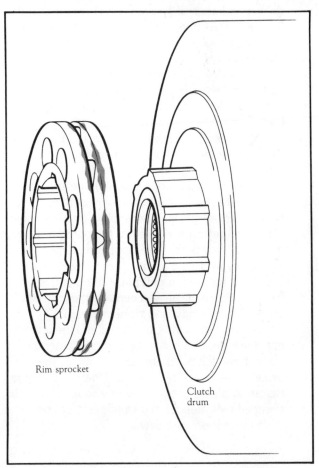

Rim sprocket

Clutch drum

Rim sprockets tend to offer better support to chain drive parts than spur sprockets do. Wear usually occurs where chain drive links engage sprocket teeth.

Should You Change a Sprocket Yourself? Some manufacturers place instructions for sprocket changing in their owner's manuals. Usually in this case, sprocket nuts are removable with an ordinary socket wrench. Other saw makers discourage sprocket removal by making the sprocket and clutch outside it removable only with special pullers or wrenches. Then, in reassembly, parts must be torqued to specifications.

In all direct-drive saws, the clutch and chain sprocket are attached to the crankshaft. If you turn the clutch, it will turn the crankshaft, which in turn moves the engine piston up and down.

So to turn the clutch nut off the shaft, you must first prevent the crankshaft from turning. You can do this in either of two ways: (1) by locking the flywheel, which the crankshaft connects directly to the clutch or (2) by going after the piston to stop its movement.

If you decide to lock the flywheel, remember that manufacturers advise that flywheel stops be specially designed for your saw. Otherwise, makeshift stops may break flywheel fins.

If you decide instead to stop piston travel, you can remove the sparkplug and stuff clothesline or starter cord through the threaded sparkplug opening to form a cushioned stop for the top of the piston as it approaches the top of its stroke. Or you can thread a special piston plug into the hole. *Caution:* Lacking special tools for clutch removal, many do-it-yourselfers attempt to break the assembly free by whacking at the outer circumference of the clutch hub with a drift pin. Although this often frees the clutch, it can also damage the crankshaft, crankshaft bearings, and crankshaft seals.

On most saws the clutch itself and the underlying clutch nut that holds the sprocket assembly in place have left-hand threads. That is, they are *usually* removed by turning them clockwise. Before applying pressure, you should check for stamped indicators on the clutch face or consult a shop manual on your saw.

With clutch and sprocket removed, check bearings for wear, replacing them if needed. Coat bearings with a water-resistant high-temperature lubricant, such as Lubriplate.

The clutch nut and the clutch itself should be retorqued to the manufacturer's specifications. For these specs, consult the manufacturer's shop manual for your saw model or else find the specs in the *Chain Saw Service Manual* published by Intertec.

Since you'll likely buy a new sprocket from a dealer anyway, you won't save much by attempting to change it yourself. Your dealer will certainly have the proper tools and knowhow.

Sprocket removal requires either simple removal of a sprocket nut or a more complex removal of the entire clutch assembly. In either case, you first need to prevent the piston from completing its upstroke. This is because the sprocket and clutch are directly driven by the crankshaft, which is connected to the piston rod. Best bet is to first screw a plastic piston plug into the sparkplug hole. Otherwise, you could stuff cotton clothesline inside the compression chamber.

III

Maintenance and Troubleshooting: Your Owner's Manual and Beyond

Why Maintenance Pays Off. Many gas-saw owners regard maintenance with displeasure, loathing, or even dread. In fact, if you cut less than a couple of cords a year, you may find that *proper* saw maintenance requires more of your time than the actual cutting. The result is often a resolve to perform little or no maintenance, using the saw sporadically—cursing it when it runs poorly and finally throwing a tantrum when it won't start.

Commonly, this leads to two options: offer the "dead" saw at a tag sale or take the saw to a service shop. Even the most successful sale will yield a big loss on the saw's purchase price. Yet if the saw was low cost to start with, costs for parts and professional labor could easily approach the saw's purchase price. If the owner decides to invest in repairs, he attends to the maintenance thereafter—consulting his owner's manual.

Sometimes professional mechanics take the trouble to explain what went wrong and why. Unfortunately, most are too busy to take a discouraged owner aside for a short course in maintenance. Then too, some are aware that poor maintenance is good for business. Sadly enough, poor maintenance and consequent mechanical problems can lead to owner injuries as well.

Even serious chainsaw users often neglect saw maintenance. In winter, especially, cold temperatures make the "gloves-off" procedures downright inconvenient. And in all seasons, sawing till nightfall gives darkness as an excuse for postponing maintenance. With the saw in storage, fuel can stratify in the fuel line and thicken in the carburetor. Also, sawdust, oil, and carbon accumulations can harden. And corrosion can take hold inside the engine and on other parts.

It's tempting to use a neglected saw a few more times on small jobs without bothering with maintenance. But before long such a saw will run poorly and may fail to start. By then maintenance tasks may have graduated to troubleshooting and major repairs.

Warranty Cautions. Upon buying a saw, pay close attention to the warranty. If you've bought from a dealer who honors warranties, you'll usually qualify for adjustments and repairs without charge for the term of the warranty. But there's a catch: If you've used improper fuel mixes or improper tools, your warranty might become void.

If you bought a warrantied saw from a dealer with no service shop, you may have to bear costs and trouble of shipping (or delivering) the saw to an authorized repair shop. These costs may outweigh those of paying for repairs locally.

Your Owner's Manual. There is no substitute for your owner's manual. In it you'll find most of what you need to know to keep your saw running well. Manuals usually cover such crucial matters as gas-oil mixtures, sparkplug specifications, carburetor adjustments, and may even show how to change the chain sprocket. Mechanical parts and procedures often differ among models by a single maker. And procedures for various saw makes may differ dramatically.

Still, general principles and procedures apply to most saws. What follows is not intended to substitute for your owner's manual. Yet it should help you in many ways.

Obtaining Service Manuals. Even if you're a mechanical wizard, you'd be wise to obtain the manufacturer's service manual on your saw before attempting tasks such as flywheel removal, ignition timing, and engine disassembly. Even clutch removal can be tricky, requiring that you know which way parts are threaded and which special tools and torques you may need. Service manuals are usually well illustrated and cover procedures even professional mechanics may need instruction on. Trouble is, service manuals are sometimes difficult to obtain. To obtain the manual for your saw, check with your service shop. For a few bucks you may get an extra manual or your service shop may be willing to order one for you. Or, if you sign your life away, you might be allowed to borrow the shop's only manual long enough to photocopy applicable pages. If all pleading fails at the shop, you might have luck requesting a service manual directly from the manufacturer. However, most saw makers don't want their service manuals in the

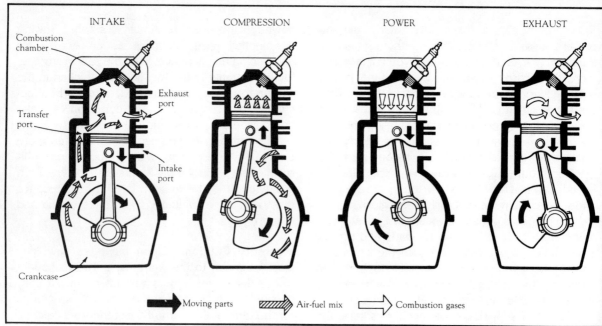

THE TWO-STROKE ENGINE CYCLE. *Unlike four-stroke cycle engines such as those in snowthrowers and most autos, two-stroke engines in chainsaws accomplish the familiar intake, compression, power, and exhaust phases with just an upward and downward stroke of the piston. Intake, power, and exhaust phases occur on the downstroke. Compression occurs on the upstroke.* **Intake:** *Nearing the bottom of its stroke, the piston covers the intake port and forces the fuel-air mixture in the crankcase through the transfer port into the combustion* chamber. (Because the exhaust port is uncovered at this moment, a small amount of the mixture escapes.) **Compression:** *The piston compresses the fuel-air mixture, while uncovering the intake port, allowing a fresh charge of fuel-air to enter the vacuum in the crankcase.* **Power:** *Just before the piston reached the top of its stroke in the compression phase, a timed spark ignited the mixture, creating this power stroke.* **Exhaust:** *The rapidly expanding gases begin to escape as soon as the descending piston uncovers the exhaust port.*

hands of "third parties," who've not attended the maker's service school.

Another option: Check a library for the latest edition of *Chain Saw Service Manual,* which includes repair tips and mechanical specifications on most saw makes and models. The *Manual* is also available from Technical Publications Division, Intertec Publishing Corp., P.O. Box 12901, Overland Park, KS 66212.

Chainsaw Engines. Gas-powered chainsaws employ a single-cylinder engine with a two-stroke cycle. In the combustion chamber, compression of fuel and then ignition occur as the piston rises. Ignition thrusts the piston downward, allowing exhaust gases to escape and forcing some of the fuel-air mixture from the crankcase into the combustion chamber. Then a new cycle begins.

Two-stroke engines are also called crankcase-scavenging engines because the fuel-air mix is scavenged from the crankcase. Here the fuel itself is a mixture of gas and oil, which serves both as fuel and engine lubricant. Two-stroke engines differ from four-strokes used in autos, snow throwers, and the larger mowers and generators. In four-strokes, the crankcase is sealed off from the combustion chamber, and crankcase oil is trapped to lubricate moving parts. Also, in most four-strokes, fuel-air mixtures enter the combustion chamber through intake and exhaust valves in the cylinder head.

At full throttle, the two-stroke engine cycle runs at anywhere from 6,000 to 11,000 rpm on most saws. That's 360,000 to 660,000 cycles per hour. A well-maintained lower-priced "consumer" saw may be able to keep up this pace for 100 to 400 hours before piston and cylinder surfaces or crankcase parts need to be renewed or changed. Heavier-duty "professional" saws may run well over 600 hours before needing such major repairs.

Unfortunately, most saws fail to survive 30 hours of use, largely owing to improper or neglectful maintenance.

FUEL

Gas-Oil Mixtures. These can greatly determine the operating life and efficiency of your engine. Two factors are vital here: (1) types of gas and oil mixed and (2) the proportion of the mix.

Oil types. Avoid dumping just any two-stroke oil into your fuel can. Snowmobiles are designed for colder operating conditions than chainsaws are. Outboards operate at more constant lower temperatures. Besides, different makes of saws and different models are designed to operate at different compressions and temperatures. So heed your manufacturer's advice implicitly on the type two-stroke oil you mix with your gas.

Gas types. For decades manufacturers recommended using only fresh *leaded* regular gasoline. Some now also recommend fresh *unleaded* gas. Fresh means that the gas is formulated for the season and hasn't begun to go stale in the fuel can. Premium-grade gases are not recommended because they burn too hot and shorten sparkplug life. Also, avoid using gasohol because it attracts moisture and hastens corrosion.

Gas-oil Ratio. This is critical. Too many saw owners simply dump some oil in with their gas, slosh the brew up, and resume cutting.

It's been proven that lower ratios of gas to oil extend the life of parts such as piston, cylinder, bearings, crankshaft, and connecting rod. This is especially so during break-in periods, which vary among manufacturers. For example, one maker recommends a 40-hour break-in period with a low ratio of 12:1 gas-oil mix before going to 16:1, using the maker's specially formulated two-cycle oil. Other makers suggest ratios of 32:1 or even 40:1, depending on the weight of the two-cycle oil.

Note: Although you want to burn a sufficiently oil-rich mixture, a too-rich mixture will foul the sparkplug early and may cause starting and running problems. Higher ratios (more gas, less oil) are of course more volatile and so lend to easier starting and increased power. But this usually means a sacrifice in lubrication as well. Still, it's far more economical to clean or replace a fouled plug at frequent intervals because of a rich mix than to replace engine parts because you burned too lean and too powerful a mix. Even when following the saw maker's recommended mix ratios, check the sparkplug electrodes at frequent intervals, comparing them to electrodes in photos in upcoming pages to analyze how the fuel is affecting them.

Compression. With proper lubrication, internal compression seals should stand up longer than the chainsawing lifetime of nonprofessional sawyers.

Yet many saws develop compression leaks either in the combustion chamber or in the crankcase before they've gotten much use. A leak in either place can cause hard starting, or rough running, or else prevent start-up altogether.

What causes loss of compression? Usually it's dirt. This may enter the engine through the air filter assembly, or it may flake off the sparkplug electrodes, or it may be drawn back in from the exhaust port.

An example of the effects of dirt? After the eruptions of Mount Saint Helens in Washington State, the volcanic ash on tree bark from Washington to Colorado would enter the fine-mesh air filters on logging saws and become a voracious lapping compound inside the engines. As a result some new saws wore out, from the inside out, in less than an hour. Volcanic ash aside, don't allow grit inside the air-filter housing while you are cleaning the filter. Also make sure the filter is sealed tight so that airborne grit can't be drawn in through edge leaks.

Again, fouled sparkplugs may also accumulate a carbon residue that can flake off and serve as an abrasive between the piston and the cylinder wall. So check frequently and clean or replace plugs as necessary.

The engine exhaust port, connected to the muffler, usually accumulates a carbon residue in time. If this builds up sufficiently, it can choke down the port opening, creating engine back pressure that reduces engine power. Similar problems result from carbon buildup in the spark arrester screen inside the muffler. More important though, a carbon buildup can allow particles to be

Periodically remove the muffler to check for buildup of carbon. Before scraping carbon, slowly pull the starter cord, moving the piston up enough to block the exhaust port, as shown. Then while you scrape, carbon particles can't fall back inside the cylinder, where they can score parts quickly. Pros usually use the rounded end of a hacksaw blade for scraping, because it's fast, and then may touch up with a wire brush. Saw manufacturers advise that you instead use a wooden scraper so that you don't scratch the piston or scratch the port area enough to hasten carbon buildup.

sucked back inside the combustion chamber. There they can score the piston and the cylinder. As the particles wear metal away, blow-by of fuel and air occurs, evidenced in loss of compression.

Dirty fuel that doesn't first plug up the fuel lines or the carburetor will damage crankshaft bearings. If the bearings become loose, the seals will fail, with consequent loss of compression. So it pays to be meticulous when refueling. *Note:* Engine overhauls (major surgery) resulting from loss of compression require the most labor and costly parts, and so cost the most. On lower-priced saws, overhaul costs for labor and parts may be greater than the saw's purchase price. It pays to ensure that the engine is properly lubricated and free of grit.

PREPARING THE SAW FOR STORAGE

Saw makers agree that you should try to avoid storing a saw for more than a few weeks. But when you must put the saw into deep storage, saw makers differ in their instruction. The chief concerns are stratification of the fuel and plain old corrosion.

Many owner's manuals recommend that you first pour all fuel out of the tank and then run the saw at idle until the fuel lines are dry. The intent here is to remove fuel that would eventually deteriorate, producing gum and other undesirable compounds. A gummed up carburetor is nobody's friend.

Trouble is, if carburetor diaphragms are left dry, they may become brittle as potato chips. Perhaps saw makers feel it's easier and less costly to replace diaphragms than to ungum a fuel system.

As an alternative to draining the fuel system, some makers may recommend adding a fuel stabilizer, such as STA-BIL, before filling the fuel tank to the top. Then they might recommend running the engine a few seconds on this mixture and stopping the engine by choking it with an overrich mixture.

Fuel stabilizers extend the time before the fuel breaks down. And the overrich mixture inside the crankcase and combustion chamber helps keep metal lubricated and free of corrosion. Yet, even fuel stabilizers may let the fuel deteriorate after a few months.

Also, filling the tank helps prevent condensation of water vapor there. The less air in the tank, the less chance of condensation problems.

If your saw maker recommends running the fuel system dry before storage, you'll probably also encounter a reminder to ward off corrosion by removing the sparkplug and pouring in a teaspoon of two-stroke oil. Then you should pull the starter rope slowly a few times to distribute the oil thoroughly before replacing the plug tightly.

It's usually best to follow your saw maker's in-structions. But it's also good to know why you are being encouraged to perform storage tasks. *Again, best bet is to fire-up your saw and operate it every few weeks, rather than storing it.* We're not all so dedicated as that, however.

Long-term Note: If you don't add fuel stabilizer for storage and if the fuel fumes could pose a fire hazard, you might be wise to drain the tank. If you do though, before start-up again, rinse the tank free of possible water condensation, using fresh gas.

Short-term Note: For storage up to a few weeks, always fill the fuel tank after use to prevent condensation.

ENGINE TROUBLESHOOTING

Most people figure that troubleshooting is a remote technology best left to guys with high mechanical aptitudes and sophisticated test equipment. True, professional mechanics can usually determine an engine's problems faster and with more assurance than the average user. A pro will also have the tools and training to undertake major surgery if necessary. Yet, you can diagnose and correct most small engine-troubles with a few basic home tools and some straight-forward logic.

There are two approaches to troubleshooting. Upcoming pages cover both.

1. *General troubleshooting.* Here you might feel your saw is running pretty well and yet suspect you can make it run better. Or the saw may just not start easily. Or you may encounter a saw at a tag sale and decide to test for troubles before taking a gamble on it.

2. *Troubleshooting specific problems.* Here you begin with a specific annoyance such as "won't start," "lacks power under a cutting load," "idles too fast," and so on.

General Troubleshooting. To start and run well, all engines need the following: (1) adequate compression in the combustion chamber, (2) ignition of the fuel-air mix at the correct time, (3) fuel and air in correct proportions. In general troubleshooting, the idea is to test each of these three areas separately and move to the next one until you find symptoms that tell you, "Here's the general trouble area." Then you attempt to pinpoint specific faults.

Upcoming pages address in sequence the compression, the ignition system, and the fuel system. This approach should give you a good understanding of the interrelationships involved. After you've gained some troubleshooting practice—especially if it's practice on one saw—you may save time by juggling the steps a bit.

Compression. An engine needs snug enough seals in its compression chamber so that there will be sufficient compression of the fuel-air mix to promote an explosive ignition force that drives the piston downward. To check compression, disconnect the sparkplug wire. Then slowly pull the starter rope, making the piston rise and then fall in-side the cylinder. If compression is good, you'll feel strong resistance to your pull as the piston approaches the top of its stroke and then a strong bounce-like release as the piston drops away.

If you don't feel the strong increase in resistance and then the bounce, the engine has poor compression. This fault may owe to one or more of the causes listed in the accompanying drawing. Repairs of these faults may mean a major overhaul that's usually best left to the pros. Yet costs for labor and parts may be so high that the investment might not be worthwhile. (*Note:* On some engines with poor compression you can peer through the exhaust port or the intake port to view piston movement. If a piston is badly scored, allowing blow-by of air and fuel, you may be able to see the

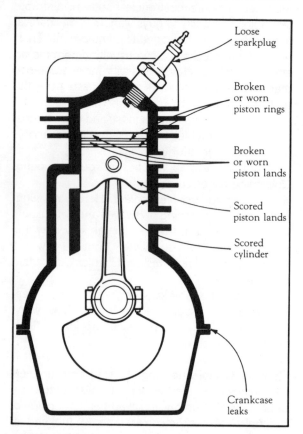

Loose sparkplug

Broken or worn piston rings

Broken or worn piston lands

Scored piston lands

Scored cylinder

Crankcase leaks

This engine cross section shows possible causes of poor compression.

An unthreaded compression gauge like this one must be forcefully held in place as you pull the starter cord. If you are alone, first remove the chain from the bar, reinstall the bar, and then clamp the bar in a sturdy bench vise to stabilize the saw while you pull. If you have a helper, the helper can stabilize the saw.

scoring. If parts are badly worn inside, you may even be able to see a slight wobble of the piston as the piston rod pushes and pulls it.)

For a conclusive test, you could rent or borrow an engine compression gauge. Some can be threaded into the sparkplug hole. Unthreaded models must be forcefully held in place. With gauge in place, pull the rope at normal starting speed up to a dozen times or until the needle on the gauge no longer rises. Most saw engines are intended to produce at least 90-100 pounds pressure per square inch (psi) at cranking speed, when cold. Larger saws may need cranking-speed compressions of 150 psi. Then compare the pressure reading with the compression figure specified for your saw by its manufacturer.

Here are possible causes for weak spark or lack of spark in a saw equipped with a mechanical breaker-point ignition system. Saws with an electronic ignition system employ a solid-state module instead of a breaker point assembly as shown. These modules can only be tested on costly equipment. Most service shops test for defects simply by substituting a new module. If that does the trick, you buy a new module—at a fairly substantial cost.

If compression feels good during the simple pull test or looks good after the pressure-gauge readout, proceed to general troubleshooting of the ignition and fuel systems.

Ignition System. The compression check you just made will give you important information on the performance of the fuel system even though you now want to concentrate on ignition. That is, the first step in checking ignition is to remove the sparkplug to examine its electrodes. If fuel is reaching the combustion chamber in response to your having pulled the starter rope, the electrodes should be wet with fresh fuel (not just a wet oil residue).

Back to the ignition system: Compare the condition of the sparkplug's electrodes with electrodes in the accompanying photos. Based on that information, you can test the plug. Or you can immediately clean it or replace it with a plug *the manufacturer has specified for your saw model.* You can clean electrodes with a special file and brush. Avoid abrasives or sandblasting, which can leave particles in the well between the center electrode and the threads that would later score engine parts. For better results though, replace the plug.

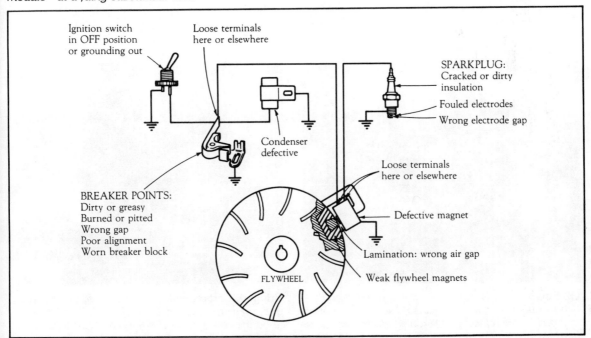

Ignition switch in OFF position or grounding out

Loose terminals here or elsewhere

SPARKPLUG: Cracked or dirty insulation

Fouled electrodes

Wrong electrode gap

Condenser defective

Loose terminals here or elsewhere

Defective magnet

BREAKER POINTS: Dirty or greasy Burned or pitted Wrong gap Poor alignment Worn breaker block

Lamination: wrong air gap

FLYWHEEL

Weak flywheel magnets

DIAGNOSING SPARKPLUG PROBLEMS

Normal. *After burning correct fuel mixes with good spark, electrodes look like this—clean, though with a light-brown color.*

Carbon Fouled. *Buildups of carbon on the electrodes can cause hard starting and poor running. Carbon can also flake off the electrodes and abrade mating surfaces of the piston and cylinder.*

Oil Fouled. *This is usually the result of an overly rich mixture (too much oil or insufficient mixing of oil and gas before refueling). Oil-fouled electrodes cause hard starting and poor performance.*

Bridged. *Carbon and oil have bridged the electrode gap, preventing spark. Lack of spark then let fuel flood the combustion chamber, as evidenced by the wet base of the plug.*

Measure the electrode gap with a wire-type gauge that matches manufacturer's specifications for your saw. This gauge reads "twenty-five thousandths" of an inch.

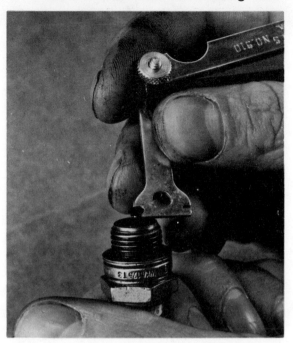

Adjust the electrode gap by prying with a special, slotted gapping tool.

Before installing any plug (new or old), ensure that the gap between its electrodes matches the saw maker's specifications for your saw model. For example, most makers specify that the gap be about .025 inch. Measure the gap with a *wire-type* sparkplug gauge (not a flat leaf gauge used for auto valve work). Procedures for gapping and spark testing are shown in accompanying photos.

If the plug you'll use in the saw gives a good spark and if your compression check was favorable, the problem probably lies in the fuel system. (*Note*: It's possible that the points on a breaker-point ignition saw are bad. Here the points could allow a strong spark that is badly timed.)

If you get no spark or a weak spark, the plug may be bad. But also inspect the wiring for faulty connections, cracks, and frayed insulation. Remedies for other possible ignition problems usually involve removal of the flywheel. This may require special tools and it always requires skillful techniques. Unless you have a service manual for your saw, flywheel removal is probably best left to the pros. Besides, further steps often require tests with electronic equipment.

To test for spark, ground the plug, as shown, on the engine or on uninsulated housing metal. Then pull the starter rope briskly. On mechanical breaker-point ignition saws, watch for a brief, snappy blue spark, as shown. On saws with electronic ignition, watch for a thin red spark, which may be hard to see unless you move the saw into the shade. Shade or no shade, you should hear the spark—like an amplified snap and crackle of Rice Krispies. Caution: Keep the plug away from the sparkplug hole; otherwise fuel spray could be ignited by the spark.

If you got no spark or a weak spark with the saw's regular sparkplug, an inexpensive test plug like the one shown can tell you if the problem lies in the original plug or elsewhere in the ignition system. If the test plug gives you a strong spark, like either of the two shown, you know that the regular plug was faulty. Replace it. If you get no spark from the test plug, the problem lies elsewhere. To consider the range of possible problems, see the ignition-system diagram, earlier. Note: You can also rig a test plug from a normal plug known to be good by breaking off the ground ("L" shaped) electrode; here you must maintain metal to metal contact of the sparkplug and the engine or its housing.

Dirty air filters are contributing causes of many engine problems. As shown here, filters vary considerably in design and filter material. Some are fine metal mesh. Others are paper. Always clean the air filter after sawing and oftener if sawdust appears to be building up, as shown on the leftmost filter.

If the spark was good, reinstall the sparkplug and proceed to general troubleshooting of the fuel system, mindful that with a mechanical breaker-point ignition, bad ignition points could be causing your troubles.

Use care when reinstalling the sparkplug. If there's a gasket above the threads, turn the plug in finger tight and then use a wrench to make about one-half turn more. If the plug has a tapered seat, with no gasket, turn it in finger tight and then *carefully* with a wrench make only about 1/32 to 1/16 turn more—just enough to make it snug.

Fuel System. After having pulled the starter rope a few times with sparkplug in place, remove the plug and examine its electrodes for traces of fuel.

If you see beads of water as well as fuel, condensation in the fuel line is causing problems. In this case it's usually wise to flush out the entire fuel system—from fuel tank to carburetor before putting the saw back to work.

If you see no fuel on the electrodes, fuel probably isn't reaching them as it should. To check this, pour about a teaspoon of fuel into the cylinder, replace the plug, and again attempt to start the engine. If the fuel hasn't been getting through until now, the engine should fire up, run a few seconds, and then expire when it has consumed the final drops of fuel.

Whether it's water or solid particles that are causing the problem, corrective steps are similar. For either problem, try the steps that follow:

1. Dump old fuel from the tank.

2. Using a hook fashioned from a wire, fish the fuel filter from the tank and examine it. If it's dirty or waterlogged, disconnect it from the fuel pickup hose. Then flush it clean or replace it.

3. If the vent for the fuel tank is clogged, the saw may begin running poorly after just a few minutes operation. If such a vent is in the fuel cap, test for clogging by loosening the cap enough to let air in and then start the saw again. If the saw now runs well, you know it's time to clean the vent.

4. Disconnect the fuel pickup hose from the carburetor and manipulate it so that fuel will drain from it. If you suspect water contamination, flush the hose by squirting gas from a small-nozzled oil

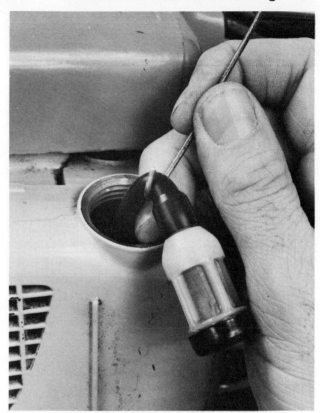

To examine the fuel filter for contamination, fish it from the tank by means of a bent wire. This filter employs metal mesh that can be flushed clean in fresh gas or disassembled and blown out. Most filters employ fabric which should be replaced periodically even if it doesn't appear to be contaminated.

can. (Pros blow compressed air through the hose to blast out contaminants and to dry the passageway.) If the hose appears to be plugged, you may be able to feel the point of blockage by squeezing the hose along its entire length. To remove blockages, remove the hose and then blow air through. Then test the passageway by using a small-nozzled oil can to squirt fresh fuel into the hose and watching for easy drainage by gravity. Inspect for contaminants in the fuel that emerges, and flush as necessary until the emerging fuel is clean. If the hose has cracks, chafed areas, bulges or soft spots, replace it. If the hose is especially soft, the cause was probably accidental mixing of bar oil with the gas.

If the carburetor fuel screen has contaminants as heavy as this, other parts of the carburetor are likely contaminated too. A service shop is best equipped to check further.

Unfortunately, many saw owners begin fiddling with carburetor adjustment screws whenever the saw doesn't start or run satisfactorily. If settings were once correct, assume they still are and perform other troubleshooting steps first. Then before messing with adjustments, study instructions in your owner's manual, which are written for your make and model carburetor.

5. Check the carburetor's vent cap to be sure it's free of gunk.

6. Remove the top cap on the carburetor and examine the fuel screen for debris. (If you earlier found the fuel filter in the tank contaminated, the carburetor screen may need cleaning also.)

7. Remove the carburetor from its connection with the pulsation passage, leading to the crankcase. To check blockage in the pulsation passage, smear a little grease over the opening and give the starter rope a few good pulls. If engine vacuum then pulses the grease in and out of the passage, you know the passage is free. If grease remains, run a wire through the passage to clear it. If your pull of the rope pulsed grease and if early troubleshooting steps showed no contaminants in the fuel hose or in the carburetor air vent or fuel screen, you've isolated the fuel system problem to the carburetor or its fuel pump. Congratulations!

8. Until now, we've not attempted to reset carburetor adjustment needles. That's partly because

many, many fuel system problems result from incorrect adjustments here. Normally, carburetor settings won't change through saw use. So if the saw once ran well on those settings, it's likely that those settings are still correct.

At this point, it's worthwhile to check the functioning of the carburetor. To do this, reconnect the dismantled parts of the fuel system, flush out the fuel tank with fresh fuel a few times to remove possible remaining water, and add a small amount of fresh fuel. Then, with sparkplug in place, and ignition on, attempt to start the saw. If it starts and runs well; you have succeeded in troubleshooting the fuel system. If the saw doesn't start, remove the plug again and inspect the electrodes. If they show no fresh fuel, the carburetor isn't letting fuel through. *Possible causes:* dirt plugging carburetor passages, gummed up fuel in passages, damaged fuel inlet needle or adjustment needles, or defective diaphragms. If diaphragms have been allowed to dry out, they may be brittle or fragmented. Car-

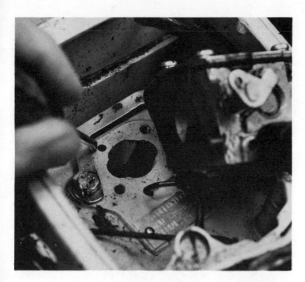

The screwdriver here points to the pulsation passage leading to the crankcase. To check for blockage, smear a little grease over the hole and, with ignition switch off, give the starter rope a few pulls. If the engine pulses the grease in and out, the passage is free. If you get no pulse, clear the passage with a wire.

buretor problems are usually best left to a pro, who can flush clogged passages with compressed air or an acid bath. If a carburetor looks to be in bad shape, pros will likely replace it rather than invest in labor and a carburetor rebuild kit. If you decide to tackle a rebuild job yourself, you may encounter only an "exploded" parts drawing with no other instructions. *Caution:* To avoid rupturing diaphragms, remove them before attempting to blow compressed air through the carburetor.

9. Carburetor Adjustments. Many operating problems result from fiddling with fuel adjustment needles. Instead of checking the compression, the air filter, or the sparkplug, many saw owners pull out that little screwdriver and go to work like a tone-deaf piano tuner.

Again, if the saw ran well at one time, it's *unlikely* that the fuel adjustment needles need adjusting, unless someone's been fiddling with them. If you're putting a factory-fresh saw into operation or if you think you ought to change adjustments,

If carburetor parts are defective, you can save the price of a new carburetor by installing kit parts. But, as this photo may imply, a complete rebuild requires the touch and patience of a jeweler. Most saw repair shops simply replace a defective part or two, rather than rebuild entirely. Otherwise, labor charges would exceed the price of a new carburetor.

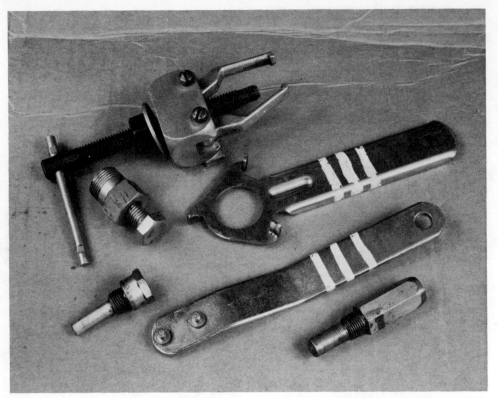

study your owner's manual. Instructions for adjustments vary significantly among carburetor makes and models.

Special tips. Begin adjustments only with a clean air filter and after you've warmed up the engine with a cutting load. To start up the engine, make preliminary carburetor settings in accord with your owner's manual. Use a soft touch, being doubly careful not to damage carburetor adjustment parts. With the saw idling, the chain should not turn.

Most saw manufactureres discourage flywheel or clutch removal without special tools. The two top leftmost tools are flywheel pullers. The two long, flat tools are clutch wrenches. And the two plugs (bottom left and right) screw into the sparkplug hole to stop piston travel so that you can turn the flywheel nut or remove the clutch. Many a clutch assembly and flywheel assembly have been damaged by over-zealous saw owners with hammers, drift pins, and pry bars.

TROUBLESHOOTING SPECIFIC FAULTS

These tables can save you money and time, while helping you become increasingly familiar with your saw. To use the tables, first locate trouble symptoms noted in the left-hand column. Then within that section, locate possible causes and remedies. Usually the simplest remedies and most common problems are listed first. Subsequent remedies in most sections tend to call for increasingly technical knowledge found in manufacturer service manuals. For remedies requiring considerable expertise, special tools, or expensive test equipment, the remedy column simply states, "Have checked at service shop."

ENGINE WON'T START, HARD TO START, CUTS OUT, MISFIRES

SYMPTOMS BY SYSTEM	POSSIBLE CAUSE	REMEDY
Fuel System	Fuel tank empty	Add fresh fuel.
	Dirty air filter	Clean or replace.
	Wrong oil-gas mix	Drain tank. Add correct mix.
	Dirty, stale, or watered fuel	Drain tank. Flush with fresh fuel. Clean, flush, or replace fuel filter. Add fresh fuel.
	Dirty or water-soaked fuel filter	Clean, flush, or replace.
	Plugged fuel hose	Clean passages to and in carburetor.
	Plugged air vents	Inspect and clear.
	Cracked, cut, or leaky fuel line	Replace.
	Kinked fuel line	Reroute, if possible, to remove kink. Replace permanently kinked line.
Carburetor fuel pump	Diaphragms damaged	Replace.
	Filter plugged	Clean or replace.
	Pulse passage plugged	Clean or replace.
Carburetor	Choke not closing	Inspect linkage and butterfly (choke plate) valve for damage. Replace if necessary.
	Adjustment needles damaged or set incorrectly	Replace and adjust, or readjust.
	Inlet valve dirty or worn	Clean or replace.
	Diaphragms or gaskets damaged	Replace.
	Wrong gasket to crankcase	Replace.
	Plugged pulse passage to crankcase	Clean with wire.
Ignition system	Switch in OFF position	Turn switch on.
	Sparkplug fouled, wrong gap, broken insulation	Clean and gap to manufacturer's specifications. Or replace with correct plug after checking gap.
	Wire insulation frayed or connection loose	Replace wire or secure connection.

Continued next page.

ENGINE WON'T START, HARD TO START, CUTS OUT, MISFIRES (Continued)

SYMPTOMS BY SYSTEM	POSSIBLE CAUSE	REMEDY
Ignition system (continued)	Incrorrect sparkplug	Replace with plug recommended by saw manufacturer, after checking gap.
	Switch grounding out	Test by disconnecting switch wire and taping aside to prevent grounding. Replace switch if necessary.
(On mechanical breaker-point ignition saws only)	Breaker points burned, dirty, wrong gap, or poor ground	Replace and regap, or clean and regap.
	Lamination: wrong air gap	Reset to specifications.
	Condenser faulty or ground poor	Tighten ground. Replace if damaged. (Low cost makes service shop test superfluous.)
	Coil damaged	Have tested at service shop.
	Flywheel magnets weak	Have tested at service shop.
(On electronic-ignition saws only)	Faulty ignition module	Have tested or replace.
Mechanical system	Sparkplug or carburetor loose	Tighten to specifications.
	Carbon buildup around exhaust port and/or in muffler	Clean away carbon with wooden scraper. (Pros use the rounded edge of a hacksaw blade, but this can scratch piston and port surfaces.)
	Crankshaft seals leaking	Replace seals.
	Flywheel key sheared	Replace key.
	Engine air leaks	Have tested at service shop.
	Rings worn, causing compression loss	Have tested at service shop.
	Engine seized, or broken connecting rod, crankshaft, etc.	Have tested at service shop.

ENGINE STARTS, RUNS BRIEFLY, STOPS

SYMPTOMS BY SYSTEM	POSSIBLE CAUSE	REMEDY
Fuel system	Fuel-tank vent plugged	Test by loosening cap. Clean or replace.
	Water in fuel mix	Drain tank. Flush with fresh fuel. Clean, flush, or replace fuel filter. Flush fuel hose and carburetor. Add fresh fuel.
	Dirty air filter	Clean or replace.
	Air leaks	Check gaskets and fuel line. Replace damaged parts.
	Fuel filter dirty	Clean or replace.
Carburetor	Wrong fuel-air mix	Adjust following instruction in owner's manual.
	Inlet valve dirty	Clean or replace.
	Diaphragms defective	Replace.
	Fuel passages clogged	Clean all passages.

ENGINE IDLES ROUGHLY

Fuel system	Dirty air filter	Clean or replace.
	Carburetor settings wrong	Refer to owner's manual.
	Damaged low-speed needle valve	Replace.
	Dirty carburetor	Disassemble and clean.
	Air leak in fuel line	Check hose for cracks and chafing. Replace if necessary. Check connections.
Mechanical	Air leak in crankcase	Have tested at service shop.

ENGINE STARVES ON ACCELERATION, IDLES TOO FAST

Fuel system	Low-speed mixture screw at wrong setting	Refer to owner's manual.
	Air leaks in carburetor or its mounting	Check gaskets and seals. Replace if necessary.
	Fuel filter dirty	Clean or replace.
	Inlet valve damaged	Replace valve and seat (if applies).
	Inlet control lever bent	Straighten or replace.

ENGINE DOES NOT RUN FULL SPEED

SYMPTOMS BY SYSTEM	POSSIBLE CAUSE	REMEDY
Carburetor	Vent plugged	Clean.
	Air or fuel filters dirty	Clean or replace.
	Choke or throttle butterfly valve not fully open	Check linkage for obstructions or damage. Check butterfly for damage.
	Inlet valve dirty	Clean and/or replace. Check fuel for contamination.
	Wrong needle adjustment	Readjust, following owner's manual. If adjustment causes no change, replace needle valve.
	Diaphragms defective	Replace diaphragms. Readjust carburetor.
Ignition	Sparkplug breaking down at high rpm	Replace sparkplug.
(On mechanical breaker-point ignition saws only)	Wrong timing	Reset.
	Breaker point gap too small	Reset.
Mechanical	Carbon buildup around exhaust port and/or in muffler	Clean away carbon with wooden scraper. (Pros use rounded edge of hacksaw blade, but this can scratch piston and port surfaces.)
	Piston rings worn or cylinder scored	Have checked by service shop.

ENGINE LACKS POWER

Fuel system	Air filter dirty (Rich mixture causes smoky exhaust.)	Clean or replace filter.
	Fuel tank not venting	Test by loosening cap. Clean or replace.
	Fuel filter or fuel lines dirty or watersoaked	Flush or replace.
	Carburetor settings wrong	Readjust following owner's manual.
Ignition system	Faulty sparkplug	Clean, check gap.
	Weak spark	On magneto-ignition saw, check gap. Reset timing. Clean or replace points.

ENGINE LACKS POWER *Continued*

SYMPTOMS BY SYSTEM	POSSIBLE CAUSE	REMEDY
Mechanical	Dull chain or improper sharpening	Sharpen chain.
	Chain brake engaged	Release brake.
	Carbon buildup around exhaust port and/or in muffler	Clean away carbon with wooden scraper. (Pros use rounded edge of hacksaw blade, but this can scratch piston and port surfaces.)
	Cylinder and/or piston scored. Piston rings worn, broken, or stuck. Crankcase seals damaged. Air leaks in crankcase or intake	Have checked at service shop.

ENGINE CONTINUES RUNNING AFTER IGNITION SWITCH IS TURNED OFF

Ignition system	Sparkplug fouled or bridged with carbon	Clean plug and check gap or replace with correct plug after checking gap.
	Faulty switch	Have checked at service shop.

ENGINE FLOODING (evidenced by unburned fuel on sparkplug or in muffler)

Fuel system	Choke closed	Open choke or check closure of choke valve.
	Vent for fuel tank plugged	Test by loosening cap. Clean or replace.
Carburetor	Fuel pump diaphragm damaged	Replace.
	Diaphragms damaged	Replace.
	Inlet valve obstructed	Check filters. Clean out carburetor.
	Inlet valve damaged	Replace valve and valve seat (if equipped).
	Inlet lever set too high	Adjust.
	Adjustment needles damaged or set incorrectly	Replace and refer to owner's manual.

ENGINE OVERHEATS (symptoms include excessive radiation of heat to your hand, stalling after prolonged use, white and blistered sparkplug electrodes)

SYMPTOMS BY SYSTEM	POSSIBLE CAUSE	REMEDY
Fuel system	Wrong fuel mix	Drain tank. Replace with fresh fuel mixed according to manufacturer recommendations.
	Intake leaks in fuel system or crankcase	Check hoses, seals, gaskets. Tighten or replace.
Carburetor	Diaphragms, inlet valve, or adjustment needles damaged	Replace.
	Too lean high-speed setting	Adjust following owner's manual.
Ignition system	Wrong sparkplug	Replace with correct plug after checking gap.
	Carbon-fouled sparkplug and cylinder	Clean, gap, and reinstall. Or replace with correct plug after gapping.
	Wrong timing	Reset.
Mechanical	Carbon buildup around exhaust port and/or in muffler	Clean away carbon with wooden scraper. (Pros use rounded edge of hacksaw blade, but this can scratch piston and port surfaces.)
	Starter housing screen clogged	Clean.
	Cylinder fins dirty	Clean.

CHAIN OILER NOT WORKING (symptoms include dry, smoking bar and chain; lack of fine oil spray from chain when saw is revved)

Mechanical system	No oil in reservoir	Replenish.
	Oil ran out too soon because of too lean winter mix with kerosene or diesel fuel	Adjust mix.
	Oil too thick in winter	Thin with kerosene or diesel fuel.
	Oil inlet holes and groove in bar clogged	Clean regularly.
	Oil outlet hole plugged	Clean.
	Oil pump, seals defective. Crankcase vents plugged.	Have checked at service shop.

CLUTCH SLIPS

SYMPTOMS BY SYSTEM	POSSIBLE CAUSE	REMEDY
Mechanical system	Chain filed incorrectly or tension too tight	Check depth gauge setting and file accordingly. Adjust chain tension.
	Chain brake engaged	Release brake.
	Clutch shoes worn or stuck	Clean clutch.

CLUTCH DRAGS, RATTLES

Mechanical system	Clutch shoes stuck, spring broken, drum out of round, shoes worn, sprocket bearing worn.	Have checked at service shop.

STARTER SLIPS

Mechanical system	Pawls worn or broken	Replace pawls and check ratchet
	Pawl spring broken	Replace.
	Ratchet worn	Replace ratchet. Check pawls for wear.

STARTER DOES NOT REWIND

Mechanical system	Spring unwound	Rewind and check for wear on rope and pulley.
	Spring broken	Replace.
	Starter jammed	Remove starter cover. Check for broken parts or foreign matter.

Sources of Mail-order Supplies

Bailey's, Inc.
P.O. Box 550
Laytonville, CA 95454

Discount logging supplies described in
seasonal catalogs and sales bulletins

Ben Meadows Company
3589 Broad St.
P.O. Box 80549
Atlanta (Chamblee), GA 30366

Forestry and logging supplies described
in large annual catalog

also

2601-B West 5th Av.
P.O. Box 2781
Eugene, OR 97402

Forestry Suppliers, Inc.
205 West Rankin St.
P.O. Box 8397
Jackson, MS 39204

Forestry and logging supplies described
in large annual catalog

Granberg Industries, Inc.
200 South Garrard Blvd.
Richmond, CA 94804

Portable chainsaw mills, file holders,
electric chainsaw sharpeners

Haddon Tool
4719 West Elm Street (Route 120)
McHenry, IL 60050

Manufacturers of the Haddon
Lumber Maker, a portable
chainsaw mill

Labonville, Inc.
R.F.D. #1
Berlin, NH 03570

Manufacturer and distributor of safety
clothing, rigging, and winches for logging

Sears Roebuck & Co.

Power Tool Catalog including chainsaws and
accessories available at Sears retail stores and
catalog sales offices

Sperber Tool Works, Inc.
Box 1224
West Caldwell, NJ 07006

Portable chainsaw mills

TSI Company
P.O. Box 151
Flanders, NJ 07836

Forestry and logging supplies described
in large annual catalog

Zip-Penn Inc.
P.O. Box 15129
Sacramento, CA 95851

Discount chainsaw parts and accessories
described in sales bulletins